W0253954

Ergebnisse der Anatomie und Entwicklungsgeschichte
Advances in Anatomy, Embryology and Cell Biology
Revues d'anatomie et de morphologie expérimentale

Springer-Verlag · Berlin · Heidelberg · New York

This journal publishes reviews and critical articles covering the entire field of normal anatomy (cytology, histology, cyto- and histochemistry, electron microscopy, macroscopy, experimental morphology and embryology and comparative anatomy). Papers dealing with anthropology and clinical morphology will also be accepted with the aim of encouraging co-operation between anatomy and related disciplines.

Papers, which may be in English, French or German, are normally commissioned, but original papers and communications may be submitted and will be considered so long as they deal with a subject comprehensively and meet the requirements of the Ergebnisse.

For speed of publication and breadth of distribution, this journal appears in single issues which can be purchased separately; 6 issues constitute one volume.

It is a fundamental condition that manuscipts submitted should not have been published elsewhere, in this or any other country, and the author must undertake not to publish elsewhere at a later date.

25 copies of each paper are supplied free of charge.

Les résultats publient des sommaires et des articles critiques concernant l'ensemble du domaine de l'anatomie normale (cytologie, histologie, cyto et histochimie, microscopie électronique, macroscopie, morphologie expérimentale, embryologie et anatomie comparée. Seront publiés en outre les articles traitant de l'anthropologie et de la morphologie clinique, en vue d'encourager la collaboration entre l'anatomie et les disciplines voisines.

Seront publiés en priorité les articles expressément demandés nous tiendrons toutefois compte des articles qui nous seront envoyés dans la mesure où ils traitent d'un sujet dans son ensemble et correspondent aux standards des «Résultats». Les publications seront faites en langues anglaise, allemande et française.

Dans l'intérêt d'une publication rapide et d'une large diffusion les travaux publiés paraitront dans des cahiers individuels, diffusés séparément: 6 cahiers forment un volume.

En principe, seuls les manuscrits qui n'ont encore été publiés ni dans le pays d'origine ni à l'étranger peuvent nous être soumis. L'auteur d'engage en outre à ne pas les publier ailleurs ultérieurement.

Les auteurs recevront 25 exemplaires gratuits de leur publication.

Die Ergebnisse dienen der Veröffentlichung zusammenfassender und kritischer Artikel aus dem Gesamtgebiet der normalen Anatomie (Cytologie, Histologie, Cyto- und Histochemie, Elektronenmikroskopie, Makroskopie, experimentelle Morphologie und Embryologie und vergleichende Anatomie). Aufgenommen werden ferner Arbeiten anthropologischen und morphologisch-klinischen Inhaltes, mit dem Ziel die Zusammenarbeit zwischen Anatomie und Nachbardisziplinen zu fördern.

Zur Veröffentlichung gelangen in erster Linie angeforderte Manuskripte, jedoch werden auch eingesandte Arbeiten und Originalmitteilungen berücksichtigt, sofern sie ein Gebiet umfassend abhandeln und den Anforderungen der „Ergebnisse" genügen. Die Veröffentlichungen erfolgen in englischer, deutscher oder französischer Sprache.

Die Arbeiten erscheinen im Interesse einer raschen Veröffentlichung und einer weiten Verbreitung als einzeln berechnete Hefte; je 6 Hefte bilden einen Band.

Grundsätzlich dürfen nur Manuskripte eingesandt werden, die vorher weder im Inland noch im Ausland veröffentlicht worden sind. Der Autor verpflichtet sich, sie auch nachträglich nicht an anderen Stellen zu publizieren.

Die Mitarbeiter erhalten von ihren Arbeiten zusammen 25 Freiexemplare.

Manuscripts should be addressed to/Envoyer les manuscrits à/Manuskripte sind zu senden an:

Prof. Dr. A. Brodal, Universitetet i Oslo, Anatomisk Institutt, Karl Johans Gate 47 (Domus Media), Oslo 1/Norwegen.

Prof. W. Hild, Department of Anatomy, The University of Texas Medical Branch, Galveston, Texas 77550 (USA).

Prof. Dr. R. Ortmann, Anatomisches Institut der Universität, 5 Köln-Lindenthal, Lindenburg.

Prof. Dr. T. H. Schiebler, Anatomisches Institut der Universität, Koellikerstraße 6, 87 Würzburg.

Prof. Dr. G. Töndury, Direktion der Anatomie, Gloriastraße 19, CH-8006 Zürich.

Prof. Dr. E. Wolff, Collège de France, Laboratoire d'Embryologie Expérimentale, 49 bis Avenue de la belle Gabrielle, Nogent-sur-Marne 94/France.

Ergebnisse der Anatomie und Entwicklungsgeschichte
Advances in Anatomy, Embryology and Cell Biology
Revues d'anatomie et de morphologie expérimentale

41 · 3

Sigmund Vaage

The Segmentation of the Primitive Neural Tube in Chick Embryos (Gallus domesticus)

A Morphological, Histochemical and Autoradiographical Investigation

With 92 Figures

Springer-Verlag Berlin Heidelberg GmbH 1969

Sigmund Vaage, M. D.
Anatomical Institute, University of Oslo
Oslo/Norway

ISBN 978-3-662-28158-1 ISBN 978-3-662-29669-1 (eBook)
DOI 10.1007/978-3-662-29669-1

Titel-Nr. 6957.

Contents

Preface

In a study of early neuroblast migration in the mesencephalon and rhombencephalon (VAAGE, 1965) it was observed that columns of neuroblasts labelled with tritiated thymidine migrated extensively in the mantle zone. In order to determine the route of migration of the neuroblasts, it was essential to define points to which the columns of neuroblasts could be referred. During the investigation a cellfree zone was observed between the mesencephalic and rhombencephalic mantles, presumably identical with the m_2-segment described by PALMGREN (1921). Several recent investigators (BENGMARK et al., 1953; BERGQUIST and KÄLLÉN, 1954), however, advocate that no morphological boundary exists between the mesencephalon and rhombencephalon.

These divergent observations and interpretations of the morphology of the neural tube in the mesencephalon and rhombencephalon prompted me to review the relevant old and recent literature concerning the morphogenesis of the neural tube. It turned out that rather divergent views on this subject are found in the literature. It was therefore deemed to be of interest to undertake a reinvestigation of the early neurogenesis. The results of this investigation are presented in this paper. As will be seen the findings concerning the neurogenesis in the chick support the opinions of several previous investigators, but in addition some new information has been obtained. Since the early neurogenesis is principally identical in chick and man the present investigation, therefore, may shed some light on the early morphogenesis in the central nervous system also in man.

I. Introduction

A. The Neuromeres

a) The Definition of the Neuromeres

In 1828 v. BAER described segmentally arranged bulges in the neural tube in chick embryos. These bulges were named "Falten" by MIHALKOVICS (1877), and neuromeres by ORR (1887).

According to the latter author, the neuromeres have the following morphological characteristics:

1. "Each neuromere is separated from its neighbours by an external dorso-ventral constriction, and opposite this an internal sharp dorso-ventral ridge, — so that each neuromere (*i.e.* one lateral half of each) appears as a small arc of a circle. The constrictions are exactly opposite on each side of the brain."

2. "The elongated cells are placed radially to the inner curved surface of the neuromere."

3. "The nuclei are generally nearer the outer surface, and approach the inner surface only towards the apex of the ridge."

4. "On the line between the apex of the internal ridge and the pit of the external depression, the cells of adjoining neuromeres are crowded together, though the cells of one neuromere do not extend into another neuromere."

(l.c., p. 335).

b) The Causes of the Formation of Neuromeres

Different opinions have been expressed concerning the causes of the neuromeres. They were found *in vivo* in salmon (HILL, 1900) and chick embryos (BERGQUIST, 1956) as well as in pig embryos dissected in amniotic fluid (STREETER, 1908). It is obvious, therefore, that the neuromeres present *in vivo* cannot be due to either the fixative agent or any other postmortal influences. A number of different *in vivo* processes were suggested as cause for their development, *e.g.*, traction produced by the flexure of the tube (FOSTER and BALFOUR, 1876; MIHALKOVICS, 1877) or by the outgrowing nerves (STREETER, 1908), pressure caused by the somites in the trunk region (NEAL, 1918), or the existence in the neural epithelium of proliferation maxima which produce localized expansions of the neural tube (BARTELMEZ, 1923). KÄLLÉN found a proliferation maximum centrally in each neuromere (1952) and a relationship between the depth of the bulges and the proliferation intensity (KÄLLÉN, 1953). A survey of the relevant literature has been given by BERGQUIST (1952a, 1956) among others.

c) The Occurrence and Number of Neuromeres

In lower vertebrates the cerebral tube is initially made up of two large bulges, a rostral and a caudal one. In amphibians and reptiles the rostral bulge, and in birds the caudal one, subdivides into two bulges (GOETTE, 1875; MIHALKOVICS, 1877; v. KUPFFER, 1906). The three bulges formed in this way are called cerebral vesicles (MALPIGHI, 1673) and conventionally named prosencephalon, mesencephalon and rhombencephalon. In fish (v. KUPFFER, 1906) and mammals (ZIEHEN, 1906) the mesencephalon originates in part from the prosencephalon and in part from the rhombencephalon. According to KUHLENBECK (1954) this holds for all vertebrates. By further subdivision of these three bulges, an increasing number of neuromeres are formed. This increase in the number of segments has been observed in both pros- and rhombencephalon (Table 1). After the closure of the neural tube, the prosencephalon consists of one bulge (MIHALKOVICS, 1877; BERANECK, 1887; MINOT, 1892; NEAL, 1898, 1918; HILL, 1900). In later stages, HIS (1888), LOCY (1895), HILL (1900) and KAMON (1906) observed two neuromeres; NEUMAYER (1899), HILL (1900) and MEEK (1907, 1909, 1910) three neuromeres; WEBER (1900) and RENDAHL (1924) four neuromeres and HOCHSTETTER (1919) five neuromeres.

In the rhombencephalon MALPIGHI (1673), v. BAER (1828) and KÖLLIKER (1861) described one bulge; BARTELMEZ (1923), ADELMANN (1925), BARTELMEZ and EVANS (1926) three neuromeres; LOCY (1895), HILL (1900), NEAL (1898, 1918) and BARTELMEZ (1923) five neuromeres; ORR (1887), McCLURE (1890), LOCY (1895), HILL (1900), ZIEHEN (1906) and v. KUPFFER (1906) six neuromeres; HOFFMANN (1889), BROMAN (1896), NEUMAYER (1899), BRADLEY (1904, 1906), THOMPSON (1907), NEAL (1918), BARTELMEZ (1923) and ADELMANN (1925) seven neuromeres; ZIMMERMANN (1891), THOMPSON (1907) and MEEK (1907, 1910) eight neuromeres.

MINOT (1892), JOHNSTON (1916) and NEAL (1918) observed that in the spinal cord the neuromeres appeared simultaneously with the somites developing in a rostrocaudal direction.

The developmental pattern of the neural tube was found to be principally the same in all vertebrates (v. BAER, 1828; KÖLLIKER, 1861; GOETTE, 1875; FOSTER and BALFOUR, 1876; MIHALKOVICS, 1877; ORR, 1887; HOFFMANN, 1889; McCLURE, 1890; ZIMMERMANN, 1891; MINOT, 1892; WATERS, 1892; LOCY, 1895; v. KUPFFER, 1895; BROMAN, 1896; NEAL, 1898, 1918; NEUMAYER, 1899; HILL, 1900; GROENBERG, 1901; BRADLEY, 1904, 1906; INGALL, 1906; KAMON, 1906; ZIEHEN, 1906; THOMPSON, 1907; MEEK, 1907, 1909, 1910; STREETER, 1908, 1911; JOHNSTON, 1909) (see also Table 1).

If the increase of the number of neuromeres takes place by further and further subdivision of the preceding ones, it may be anticipated that there is a continuous increase of neuromeres during the ontogenesis of the brain tube. The varying number of neuromeres observed may, therefore, be explained as due to examination of different ontogenetic stages.

BERGQUIST, KÄLLÉN and their collaborators in a series of publications have materially widened our knowledge of the early developmental pattern of the vertebrate brain. Previous results are confirmed by these investigations, showing that the cerebral neuromeres are homologous in all vertebrates (BERGQUIST,

Table 1. *Survey of the number of bulges observed in the prosencephalon, mesencephalon and rhombencephalon during neurogenesis*

		P	M	R
	Mammals			
MIHALKOVICS (1877)	Rabbit	2	1	5 (?)
v. KUPFFER (1885)	Man, Sheep, Mouse	3	2	6
ZIMMERMANN (1891)	Rabbit	2	3	8
BROMAN (1896)	Man			7
NEUMAYER (1899)	Sheep, (rabbit)	3	2	7
BRADLEY (1904, 1906)	Pig			7
ZIEHEN (1906)	Man, etc.	3	1	6
INGALL (1906)	Man			7
THOMPSON (1907)	Man	2	2	7 (8?)
STREETER (1908, 1911)	Pig	2	1	6
MEEK (1910)	Rabbit	3	2	8
HOCHSTETTER (1919)	Man	5	1	8
PALMGREN (1921)	Mouse	3	2	
BARTELMEZ (1923)	Man	2	2	7
ADELMANN (1925)	Rat		2	7
BARTELMEZ and EVANS (1926)	Man	2	2	7
BERGQUIST and KÄLLÉN (1953a, b, 1954)	Man, etc.	4	2	6
	Birds			
MALPIGHI (1673)	Chick	1	1	1
v. BAER (1828)	Chick	2	1	2
FOSTER and BALFOUR (1876)	Chick	2	1	4 (?)
MIHALKOVICS (1877)	Chick	2	1	5 (?)
McCLURE (1887)	Chick	4	2	6
ZIMMERMANN (1891)	Chick	2	3	8
HILL (1900)	Chick	3	2	6
WEBER (1900)	Fasan	2	2	
v. KUPFFER (1906)	Chick, Sparrow	3	2	6
KAMON (1906)	Chick	3	2	7
MEEK (1907)	Pheasant	3	2	8
NEAL (1918)	Chick	1	1	7 (5?)
PALMGREN (1921)	Chick	3	2	
RENDAHL (1924)	Chick	5	2	
STREETER (1933)	Chick	2	1	6
BERGQUIST and KÄLLÉN (1953a, b, 1954)	Several	4	2	6
	Reptiles			
ORR (1887)	Lizard	3	1	6
HOFFMANN (1889)	Lizard, Tropidon			7
McCLURE (1890)	Lizard	4	2	6
v. KUPFFER (1906)	Lizard, Anguis	3	2	6
PALMGREN (1921)	Tropidonotus	3	2	
BERGQUIST and KÄLLÉN (1953a, b, 1954)	Several	4	2	6
	Amphibians			
GOETTE (1875)	Bombinator	1	1	1
v. KUPFFER (1895, 1906)	Necturus, Rana Salamandra	3	2	6
McCLURE (1890)	Ambystoma	4	2	5

Table 1 (Continued)

		P	M	R
Waters (1892)	Ambystoma	3	2	5
Palmgren (1921)	Rana	3	2	
Bergquist and Källén (1953a, b, 1954)	Several	4	2	6
	Fish			
Ahlborn (1883)	Petromyzon	2	1	2 (?)
Scott (1887)	Petromyzon	2	2	
Zimmermann (1891)	Acanthias, Mustulus	2	3	8
Waters (1892)	Gadus (cod)	3	2	6
Locy (1895)	Squalus	3	2	6
Neal (1892)	Squalus	1	1	7 (5 ?)
Hill (1900)	Salmon	3	2	6
v. Kupffer (1906)	Acanthias, Acipenser Salmon	3	2	6
Meek (1909)	Acanthias	3	2	8
Palmgren (1921)	Squalus, Salmon	3	2	
Bergquist and Källén (1953a, b, 1954)	Several	4	2	6

1952a, b) and that the rostral and the caudal encephalomeres are serially homologous (Källén, 1954).

However, the view advocated by Bergquist and Källén is opposed to the idea of a continuous subdivision of the neuromeres. Thus Källén and Lindskog (1953) describe the segmentation of the neural tube in Mus as two consecutive rostrocaudal waves of segmentation. The first wave, called the proneuromeric wave, disappears in a caudorostral direction before the next, the neuromeric wave, commences. The neuromeres likewise disappear completely before a third, the postneuromeric wave of Bergquist and Källén (1954), occurs. Similar waves of segmentation are found in other vertebrates as well (Wedin, 1954; Bergquist and Källén, 1954, 1955).

According to this concept of the segmentation pattern, there are six (or five) proneuromeres, eleven (or nine) neuromeres and thirteen postneuromeres. In the rhombencephalon Bergquist and Källén (1953a, b, 1954, 1955) describe three prorhombomeres, six rhombomeres and six postrhombomeres.

Hence there are two different points of view concerning both the segmentation pattern of the neural tube and the number of neuromeres. According to one view, an increasing number of neuromeres are formed by successive subdivisions of the preceding ones. According to the view advocated by Bergquist and Källén, on the other hand, three different series of neuromeres occur.

d) The Homology of Neuromeres

In the literature a great effort is made to define homologous segments (neuromeres) in various vertebrates. Homologous neuromeres are neuromeres which in different vertebrates originate from the identical components of the neural tube and then develop in a similar way for a longer or shorter time (Källén, 1959). Different views are still pending. This may be illustrated by three examples.

1. The second bulge in the primitive brain was called mesencephalon by Malpighi (1673), v. Baer (1828), Kölliker (1861), His (1868), Hill (1900), Neal (1918) and Patten (1958) among others. The same bulge in a comparable developmental stage of chick embryos was named metencephalon by Beraneck (1887) and Bergquist (1956).

2. After the subdivision of the prosencephalon, two bulges are present in the forebrain. The caudal one represents the anlage of the diencephalon (parencephalon and synencephalon) according to v. Baer (1828), Kölliker (1861), His (1868), Hill (1900) and others, while Bergquist, Källén and their collaborators maintain that the same bulge represents the anlage to the synencephalon only (Bergquist and Källén, 1954, 1955).

3. A third example is the otic rhombomere which according to Orr (Table 1) does not connect with any sensory nerve. Meek (1907, 1909, 1910) described two rhombomeres at that particular level of the neural tube. v. Kupffer (1906) and Bergquist and Källén (1954) observed no nerve-free neuromere in the same region. Differences in the descriptions of the rhombomere in question occur even in the same textbook. Thus in „Handbuch der vergleichenden und experimentellen Entwicklungslehre der Wirbeltiere" (Hertwig, 1906) v. Kupffer interpreted the otic rhombomere as more recently Bergquist and Källén do, while Ziehen concurred with Orr.

The controversies still pending make it appropriate, therefore, to re-investigate the problem.

B. The Relationship between the Cranial Nerves and the Neuromeres

The relationship between the neuromeres and the cranial nerves is divergently described in the literature. The nerves differentiate from the neural crest (Marshall, 1878; Goronowitsch, 1893; Lillie, 1919) located along the dorsal surface of the neural tube. The neural crest located dorsolateral to the second, fourth, sixth and seventh rhombomeres differentiates into sensory nerves according to Orr (1887), Hoffmann (1889), Zimmermann (1891), Waters (1892), Locy (1895), Broman (1896), Neumayer (1899), Bradley (1904, 1906), Ingall (1906), Ziehen (1906) and Thompson (1907). Hill (1900) described the pattern in this manner: "From segment 7 the main fibres of the trigeminal nerve pass cephalad to the Gasserian ganglion (Figs. 49 and 50)[1]. Segment 8 has no nerve connection. Segment 9 gives rise to the fibres of the seventh and eighth pair of nerves (Figs. 49 and 50, VII and VIII). Segment 11 is connected with the fibers of the ninth pair of cranial nerves (Fig. 50)" (*l.c.*, p. 424). According to the aforementioned authors, the fifth rhombomere, called the otic one, gives off no nerve, but the space lateral to it is occupied by the otic vesicle. Some authors, however, do not describe any nerve-free otic rhombomere (v. Kupffer, 1906; Bergquist and Källén, 1953a, 1954), while Meek (1907, 1910) described two nervefree rhombomeres in the otic region. The oculomotor nerve is said to originate from the rostral mesencephalic segment by Zimmermann (1891), from the caudal mesencephalic segment by Bergquist and Källén (1954, 1955) and from the rostral rhombencephalic segment by Ahlborn (1883).

1. Figure numbers mentioned in this quotation refer to Hill's figures.

According to His (1888), Hoffmann (1889), Neumayer (1899), Kamon (1906), Ziehen (1906) and Hochstetter (1919) the trochlear nerve develops within the rostral rhombencephalic segment, while Zimmermann (1891), Waters (1892), Hill (1900) and Hugosson (1957) maintain that the trochlear nerve derives from the caudal mesencephalic segment. The abducent nerve originates from the third rhombomere (Hill, 1900).

Controversial conceptions of the relationship between the neural tube and the cranial nerves hardly can be due to species differences because in several instances the interpretations were based on investigations of embryos of a variety of vertebrates (Table 1). Some of the disagreements at least are most likely due to misinterpretation of the homology of the bulges in different vertebrates. Obviously further investigations are indispensable in order to reveal the relationship between the neuromeres and the cranial nerves.

C. The Relationships between the Cerebral Divisions and the Neuromeres

After its closure the neural tube can be subdivided into a rostral, cerebral, and a caudal, spinal part (McClure, 1890). In all vertebrates, the former can be subdivided into three encephalomeres, *viz.*, the pros-, mes- and rhombencephalon (see p. 9). The boundaries between these subdivisions persist during ontogenesis, separating the pros-mesencephalon, the mes-rhombencephalon and the rhombencephalon and the spinal tube. The boundaries are, however, variously placed by different authors.

a) The Pros-Mesencephalic Boundary

It was placed immediately caudal to the posterior commissure by Neal (1898). Orr (1887), Haller (1929), Herrick (1948) and Kuhlenbeck (1954), on the other hand, placed the boundary rostral to the posterior commissure.

b) The Mes-Rhombencephalic Boundary

Beraneck (1887) and Bergquist (1956) apparently incorporated the mesencephalic bulge in the metencephalon in early stages. The mes-rhombencephalic boundary was therefore identified by them with the pros-mesencephalic boundary. Ahlborn (1883) suggested that in Petromyzon the mesencephalon did not comprise any floor region. The mes-rhombencephalic boundary, therefore, was described as coinciding ventrally with the di-mesencephalic boundary immediately rostral to the oculomotor nucleus. By studying very closely graded stages of embryos of different species, His (1880, 1888), Hoffmann (1889), Zimmermann (1891), Broman (1896), Neumayer (1899), Bradley (1904, 1906), v. Kupffer (1906), Thompson (1907), Neal (1918) and Palmgren (1921) were led to place the boundary immediately rostral to the trochlear nucleus, while Waters (1892), Locy (1895) and Hill (1900) placed it caudal to the same nucleus, and McClure (1890) and Hugosson (1957) placed it still further caudal in the neural tube. It has proved difficult so far to find out which one of these boundaries coincides with the primitive mes-rhombencephalic boundary.

The divergent opinions concerning the boundary in question can hardly be due to species differences. More likely the explanation is to be found in differing interpretations of the observations.

c) The Rhombo-Spinal Boundary

v. Baer (1828), Mahalkovics (1877) and McClure (1890) placed the boundary at the level with the first pair of metameric somites. Other authors, using different criteria for determination of the boundary, placed it at different levels (Meek, 1907, 1909, 1910; Johnston, 1916; Bergquist, 1952a). So far it has proved difficult to define the primitive rhombo-spinal boundary.

d) The Tel-Diencephalic Boundary

The two bulges of the prosencephalon are separated dorsally by a conspicuous furrow which is the anlage to the transverse septum. However, the tel-diencephalic boundary is variously placed by the different investigators.

The telencephalon (prosencephalon secundarium), according to Mihalkovics (1877, p. 30—31), is formed by dorsorostral expansion of the lateral wall of the prosencephalon (prosencephalon primitivum). „Die Abgrenzung der neu entstandenen Blase gegen das primäre Vorderhirn ist anfangs eine sehr schwache, in zwei unbestimmten Einschnürungen der Seitenwände bestehend, welche vor und über der Einmündung der Sehnervenstiele im schwachen Bogen nach oben verlaufen". „Die Grenzen kommen erst nach der vollständigen Ausstülpung des secundären Vorderhirns schärfer zum Vorschein, doch ist die Abgrenzung am Boden auch dann noch keine bestimmt markierte und muß künstlich vor die Sehnervenplatte gestreckt werden." Dorsally this boundary extends from the transverse system, and ventrally it terminates rostral to the optic recess. A homologous boundary zone was observed in fish, birds and mammals (Mihalkovics, 1877; Meek, 1907, 1909, 1910; Kuhlenbeck, 1954, 1956).

Other authors showed that the prosencephalon subdivides into two bulges separated by a ventricular ridge. This begins dorsally in the transverse septum, extends laterally and ventrally and terminates between the infundibular anlage and the mammillary recess (His, 1888; Neumayer, 1899; Weber, 1900; Kamon, 1906; Ziehen, 1906).

In accordance with this observation the rudiments of the eyes, the infundibulum and hemispheres are situated rostral to the tel-diencephalic sulcus.

Still an other author (v. Kupffer, 1906) described a ventricular tel-diencephalic boundary sulcus which extends dorsoventrally from the transverse septum to the optic recess. The infundibular region and the optic chiasma belong accordingly to the diencephalon. A homologous boundary zone was observed in fish, amphibians, reptiles and birds (v. Kupffer, 1895, 1906; Kuhlenbeck, 1954).

D. The Relationship between Several Secondary Structures and the Neuromeres

For want of a distinct morphological boundary between the mesencephalon and the rhombencephalon various secondary structures have been introduced to define the border.

The structures most frequently used are the notochord (Ahlborn, 1883), the floor plate (Kingsbury, 1920), and the fovea isthmi (Herrick, 1917). Some of these structures and the distribution of the glycogen content in the floor of the neural tube (Kingsbury, 1934) are used conventionally without attempts being made to define their relationships to the neuromeric pattern.

In order to determine the relationship between the neuromeres and the interneuromeric boundaries on the one hand and the notochord, the floor plate, the fovea isthmi, the posterior intraencephalic sulcus and the sulcus isthmi on the other hand, a re-examination based on closely graded developmental stages seems indispensable.

E. The Aim of the Present Investigation

The divergent views concerning the developmental pattern and the subdivision of the neural tube evidently are independent of the vertebrate species investigated. The logical step, therefore, in order to clear up the controversies that are pending, is to make a detailed investigation of the ontogenetic development of the neural tube in one species. Hence the present investigation is limited to *Gallus domesticus* of which closely graded stages are obtainable in desirable numbers. The investigation is restricted to the developmental period between the time of closure of the neural tube (*HH–8*) and the establishment of a permanent pattern at the beginning of the sixth day (*HH–27*).

The purpose of this research is to investigate:

1. The developmental pattern of the neural tube.
2. The relationship between the nerves III–X and the neuromeres.
3. The relationships between the various intracerebral boundaries and the neuromeres during the ontogenesis of the neural tube.
4. The relationships between the notochord, the floor plate, the fovea isthmi, the sulcus intraencephalicus posterior and the distribution of the glycogen in the floor of the neural tube and the subdivisions of the neural tube.

The bulges in the neural tube are named in accordance with the nomenclature introduced by v. Baer (1828), Orr (1887), McClure (1890), Meek (1907) and Bartelmez (1923) as shown in Table 2.

Table 2. *Correlation between the brain regions and the designations of the neural bulges*

<table>
<tr><td colspan="3">Encephalon (brain tube)</td><td rowspan="3">Myelon (spinal cord)</td></tr>
<tr><td colspan="3">Encephalomeres</td></tr>
<tr><td>Prosencephalon
P</td><td>Mesencephalon
M</td><td>Rhombencephalon
R</td></tr>
<tr><td>Prosomeres</td><td>Mesomeres</td><td>Rhombomeres</td><td>Myleomeres</td></tr>
<tr><td colspan="4">Neuromeres</td></tr>
</table>

The criteria introduced by Orr (1887) are used for definition of the neuromeres (see p. 8).

It has proved unsuitable to use names as the telencephalon, diencephalon, thalamencephalon, proneuromeres and postneuromeres, since these names have been employed to designate regions that are not homologous (see Discussion). In Table 3 I have listed the different bulges and the synonyms used in the anatomical literature.

Table 3. *Correlation between the brain regions and the terms used by* NEUMAYER (1899), MEEK (1907), HOCHSTETTER (1919), BARTELMEZ (1923), RENDAHL (1924), BERGQUIST and KÄLLÉN (1954) *and the author*

	NEUMAYER (1899)	MEEK (1907)	HOCHSTETTER (1919)	BARTELMEZ (1923)	RENDAHL (1924)	BERGQUIST and KÄLLÉN (1954)	VAAGE
Telencephalon	T	P_1	T	T	T	I, II } 1,2	pr_1, pr_2
Sulcus i.a			N_1				pr_3
anterior portion			N_2		P_1	III { 3	pr_5
Parencephalon	P	P_2		D			
posterior portion			N_3		P_2	4	pr_6
precommissural a.					S_a		pr_7
Synencephalon	S	P_3				IV 5	
commissural area			Co.p.		S_b		pr_8
Mesencephalon	m_1, m_2	m_1, m_2	M	m_1, m_2		V 6, 7	m_1, m_2
Isthmus rhombencephali			Is				Is
First rhombomere	1	Rh 1	n_1	Rh 1		VI 8	rh_1
Second rhombomere	2	Rh 2	n_2	Rh 2		VII 9	rh_2
Third rhombomere	3	Rh 3	n_3	Rh 3		VIII 10	rh_3
Fourth rhombomere	4	Rh 4	n_4	Rh 4		IX 11	rh_4
Fifth rhombomere	5	Rh 5, Rh 6	n_5	Rh 5			rh_5
Sixth rhombomere	6	Rh 7	n_6	Rh 6		X 12	rh_6
Seventh rhombomere	7	Rh 8	n_7	Rh 7		XI 13	rh_7

II. Material and Methods

Chick (*Gallus domesticus*) embryos were used in this investigation. The eggs were incubated at 38° C. The embryos were as a rule fixed in Carnoy solution and graded according to the stages of HAMBURGER and HAMILTON (1951). The stage number is placed as a suffix after HH⁻. After dehydration in alcohol and embedding in paraffin serial sections were prepared in the three conventional planes at 10—20 μ. The sections were stained with iron-hematoxylin or hematoxyline-eosine. A few embryos were fixed for control in 10% formalin or Bouin's solution. Graphical reconstructions and wax models were made of the neural tube at the different developmental stages. The embryos used for reconstructions were coembedded with one to three sharp-edged slices of tissue (*c′*, *c″*, *c‴*, Fig. 1) which were placed vertically (Fig. 1a) to the plane of sectioning (*d*—*d*, Fig. 1a).

The reconstructions (and photographs) were checked by examination of the outer and inner surfaces of the neural tube in unsectioned embryos under a binocular stereomicroscope. Having carefully dissected away the mesoderm under a binocular microscope using small metal rods, 30 μ in diameter, the outer surface was exposed. The bulges of the tube and the nerves could then be easily observed. Some of the objects were photographed. The inner surface was exposed in quite another way. The objects (which were embedded in paraffin in the usual way) were sectioned in one of the three conventional planes until an appropriate

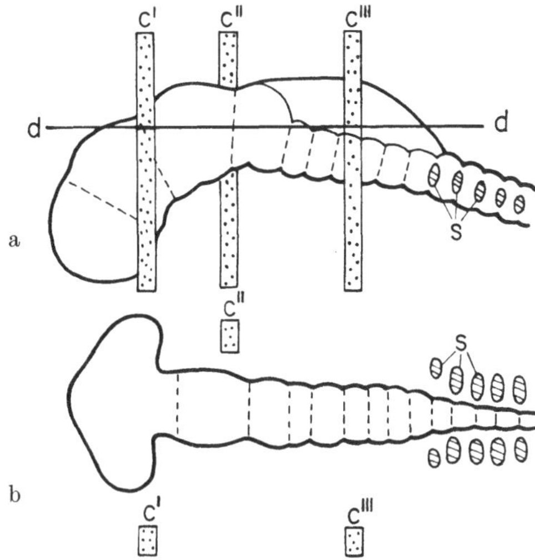

Fig. 1a und b. Diagrammatic drawing of lateral (a) and dorsal (b) view of an embryo coembedded with three sharp-edged tissue slices (*c'*, *c''* and *c'''*). The slices are placed at right angles to the appropriate sectioning plan (*d–d*)

Abbreviations for all figures

A	archencephalon
a.dm	dorsal mesencepahlic area
a.vm	ventral mesencephalic area
Bu	bucco-pharyngeal membrane
C	cerebellum
Ca	caudal
C.f	cranial flexure
Ch	notochord
Cha	chiasma opticum
Cr.f	cervical flexure
D	deuteroencephalon
d	ductulus opticus
E.mp	posterior mesencephalic eminence
Ep	epiphysis
Ep.nt	epichordal neural tube
F	interneuromeric furrow
F.A–D	arch-deuteroencephalic boundary and furrow
F.i	isthmic fovea
Fl.pl	floor plate
$F.B_1$–B_2	PrB_1–PrB_2–furrow
F.M	Fissura Monroi
F.m–r	mes-rhombencephalic boundary and furrow
F.my–sp	myelo-spinal boundary and furrow
F.p–m	pros-mesencephalic boundary and furrow
F.r–sp	rhombo-spinal boundary and furrow
$F.t$–d_1	primary tel-diencephalic boundary and furrow
$F.t$–d_2	secondary tel-diencephalic boundary and furrow
g	glycogen-containing raphe
$G_V, G_{VII-VIII}$ ...	ganglion of nerves V, VII–VIII, ...
HH	stage according to the staging of Hamburger and Hamilton (1951)
Hy	hypophysis
Is, Is_a, Is_b	rhombencephalic segments Is, Is_a, Is_b
I.s.c 1, ...	intersomitic cleft 1, ...
Is.m	isthmic migration
L.s	left side
l.s	longitudinal sulcus
M	mesencephalon
Ma	mammillary recess
m.m	mantle layer migration
my	myelomere
$my_1, my_2, \ldots$	myelomere 1, 2, ...
m_1, m_2	mesomere 1, 2
Na	anterior neuropore
N.C. 1	first cervical nerve
N.f	neural furrow
N.op	nervus opticus
N.p	neural plate

Abbreviations for all figures (Continued)

Abbreviation	Meaning
NPC	non-proliferating cell column
N.t	neural tube
Nucl. III, IV, ...	nucleus nervus III, IV, ...
Ol.p	olfactory pit
Op.e	optic evagination
Ot.p	otic placode
Ot.v	otic vesicle
P	prosencephalon
p.a	primary area
P–R.nc	pros-rhombencephalic division of the neural crest
Ph.a 1, Ph.a 2, ...	pharyngeal arch 1, 2, ...
Ph.g 1, Ph.g 2, ...	pharyngeal groove 1, 2, ...
Po.nc	postotic division of the neural crest
PrA, PrA_1	prosomere A, A_1 ...
pr_1, pr_2	prosomere 1, 2 ...
pr.ch	prechordal plate
Pr.nc	preotic division of the neural crest
Pr.nt	prechordal neural tube
Prol.a	proliferative activity
R	rhombencephalon
r	interneuromeric ridge
r.rh	raphe rhombencephali
Ra	Ratches pouch
RhA, RhB, ...	rhombomere A, B ...
rh_1, rh_2 ...	rhombomere 1, 2 ...
"rh_7"	"rhombomere 7"
R.i	recessus infundibuli
r.l	rhombic lip
R.m.r	relative mitotic rete
r.o	recessus opticus
r.p	roof plate
R. "pr_{4-5}"/pr_6	internevromeric ridge "pr_{4-5}"/pr_6
Ros	rostral
R.s	right side
S, S_1, S_2 ...	somite, somite 1, somite 2 ...
Se	Sessels pocket
Seg	segment
S.dl	sulcus dorsolateralis
S.i.a	sulcus intraencephalicus anterior
S.i.p	sulcus intraencephalicus posterior
S.m	sulcus longitudinalis medialis
S.l	sulcus limitans
S.l.Met	sulcus lateralis metencephali
Sp	spherencephalon
Sp.c	spinal cord
Sp.nc	spinal division of the neural crest
S.pr	prosencephalic sulcus
S.rh	rhombencephalic sulcus
s.t	septum transversum
su	neuromeric sulcus
S.vl	sulcus ventrolateralis
Sy.z	synthetic zone
T	telencephalon
t.p	tuberculum posterior

level was obtained and thereafter re-transferred into absolute alcohol. The morphological patterns on the inner surface could then be examined. Some of the objects were photographed. Subsequently they were re-embedded in paraffin and serially sectioned. The photos could then be checked by means of the histological sections.

Some embryos were dissected in the amniotic fluid. Others were stained *in vivo* with neutral red stain dissolved in physiological saline (0.9% NaCl). Some embryos were allowed to survive for more than 2 days and they had a normal neural differentiation. It was possible, with a binocular stereomicroscope, to observe the same bulges in the neural tube as in the fixed embryos. The possibility could thus be excluded that the bulges, present on the outer and inner surfaces of the neural tube, had been caused by the fixative.

The distribution of the glycogen in the "floor plate" was studied in embryos fixed in alcoholic formaldehyde (4% HCHO in 85% ethanol neutralized in $CaCO_3$) for 3—4 hours, embedded in paraffin, sectioned serially at 15 μ, mounted and oxidized in periodic acid for 10 minutes. Control sections were incubated for 40 minutes in 1% amylase dissolved in physiological saline solution, and thereafter oxidized according to the same procedure as above. SCHIFF's procedure was performed according to a technique modified by LILLIE (PEARSE, 1960).

The proliferating ability of the cells in the neural epithelium was studied by injection of labelled thymidine (15—25 micro-Curie) in the yolk sac. The injection was performed through a window in the egg shell, opened and closed according to the method of HARKMARK and GRAHAM (1951). The eggs were re-incubated for 30—60 minutes or more. The embryos were fixed in CARNOY'S solution, embedded in paraffin, sectioned serially at 4—5 μ, and autoradiographed according to the dipping method (MESSIER and LEBLOND, 1957). Kodak NTB-2 emulsion was used. Control sections were incubated for 30 minutes at 37° C in an M/40 acetate-veronal buffer of pH 7.5 containing 1 mg/100 highly purified pancreatic desoxyribonuclease (SANDERS, 1946), before being autoradiographed. The sections were exposed in black PVC boxes at 4° Celcius for 2—10 days. They were then developed, fixed and stained with toluidine or hematoxylin-eosine.

III. Observations on Living Embryos

The procedure in question makes it possible to investigate the formation of some bulges in the neural tube during early ontogenesis. The following describes the bulges observed in living embryos between the stages *HH–8–9* and *HH–16*.

Results

Stage HH–8–9. The embryo is in the early neural tube stage (Fig. 2). Rostrally the neural folds have converged, forming a neural tube. More caudally the neural folds diverge and form a furrow. The rostral somite, which is indistinctly separated from the head mesoderm, is situated cranial to the first intersomitic cleft.

Stage HH–9. The coalition of the neural folds has proceeded caudalwards (Fig. 3). Dorsally the line of fusion can be seen extending between a wide open anterior neuropore and a rhomboidal sinus. On both sides of the caudal half of the tube, seven distinct somites have developed (not all shown). The rostral one is, as in the preceding stage, situated cranial to the first intersomitic cleft. Coagulation (electro-diathermi) and labelling (with coal label) experiments prove that the rostral somite is identical with the rostral one present at the preceding stage (VAAGE and HØIVIK, 1969). Medial to the rostral somite, a superficial furrow (*F.r–sp*) sibdivides the prospective central nervous system into a cerebral and a spinal tube. The cerebral tube is further subdivided into a short broad rostral and a long narrow caudal encephalomere. The two encephalomeres are presumably identical with the archencephalon (*A*) and the deuteroencephalon (*D*), respeetively, described by v. KUPFFER (1906), among others. The prospective spinal cord consists of several successive bulges calles myelomeres by MCCLURE (1890). Each myelomere is interposed between two succeeding pairs of somites. The rostral and the caudal myelomeres are situated medial to the first and the sixth (not shown in the figure) intersomitic cleft, respectively.

Stage HH–10. The neural tube is further transformed (Figs. 4—6). The archencephalon (prosencephalon) has shortened rostrocaudally, but it is broader compared with the preceding stage. The deuteroencephalon is subdivided into a rostral called mesencephalon (*M*) and a long narrow caudal called rhombencephalon (*R*). Coal labelling experiments (VAAGE and HØIVIK, 1969), prove that the pros-mesencephalic boundary at this and later stages is identical with the arch-deuteroencephalic boundary at stage *HH–9.* In the rhombencephalon, at least three rhombomeres (*RhA*, *RhB* and *RhC*) are distinguishable. These are probably homologous to the three primary rhombomeres described in mammals by BARTELMEZ (1923). The rostral rhombomere is faintly subdivided into a rostral (RhA_1) and a caudal (rh_3) one. In the spinal cord, three new myelomeres are added to

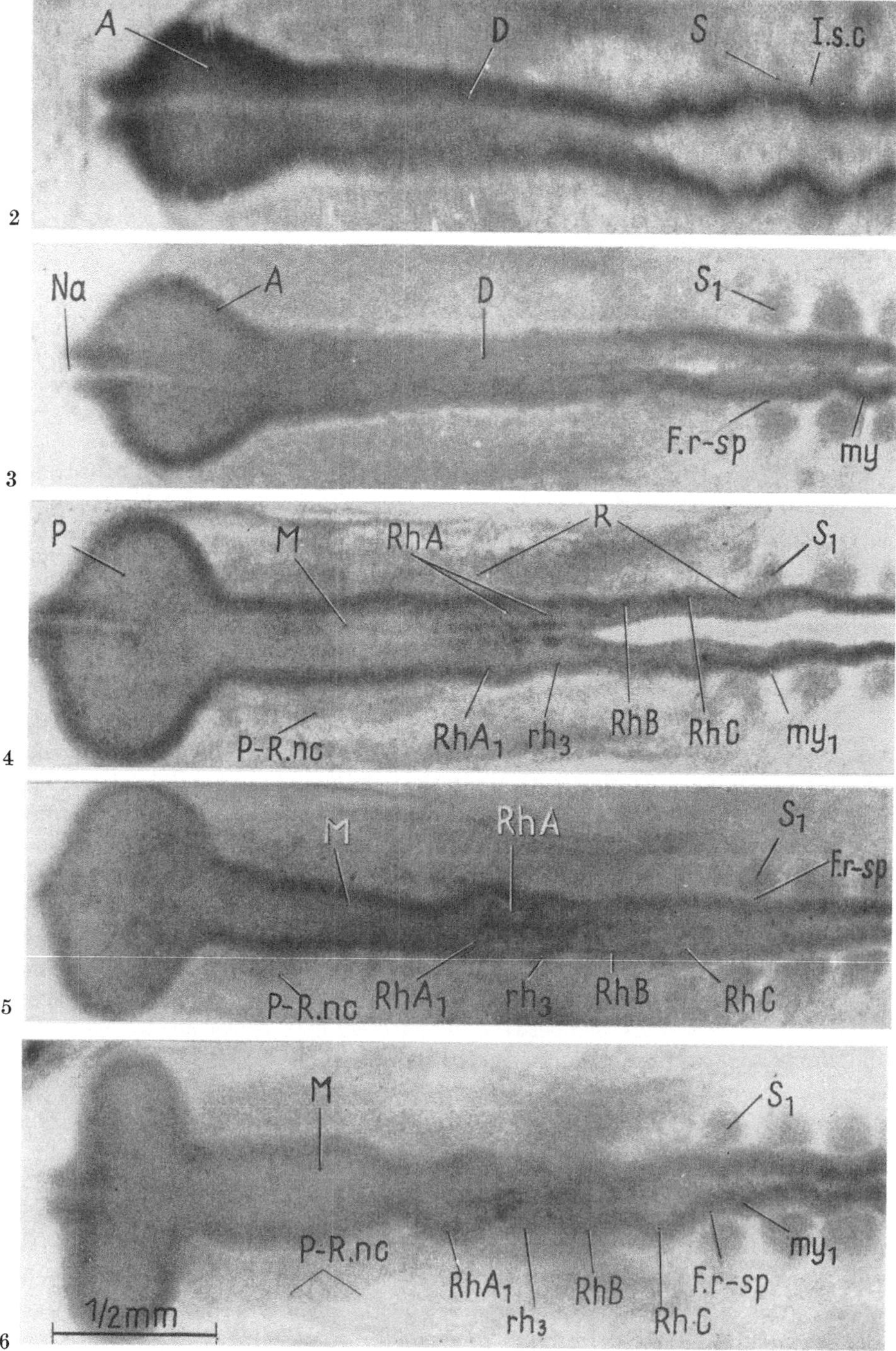

Figs. 2—6. Dorsal views of the cerebral tube and the rostralmost portion of the spinal tube of living chick embryos at stages *HH–8–9*, *HH–9*, *HH–9–10*, *HH–10*, *HH–10+*

the six present at stage *HH–9*. Lateral to the spinal cord are ten distinct somites. The rostral one is situated cranially to the first intersomitic cleft. Coagulation and labelling experiments prove that the rostral somite is identical with the rostral one present at the preceding stages (VAAGE and HØIVIK, 1969).

Along the lateral border of the mesencephalon and the caudal portion of the prosencephalon an opaque column is observed from the stage *HH–9–10* (Fig. 4) onwards. This structure evidently represents the prospective pros-mesencephalic neural crest described by LILLIE (1919) and HOLMDAHL (1928). Shortly later (*HH–10*) the neural crest anlage has grown further laterally and caudally and terminates lateral to the *RhA* (Fig. 5). At stage *HH–10+* the lateral border of the neural crest may be seen to approach the lateral boundary of the head mesoderm (Fig. 6).

Stage HH–11. The bulges of the neural tube are still more distinct (Fig. 7). The prosencephalon is bent ventrally and subdivided into two bulges. The caudal bulge (*PrB*) has a width similar to that of the mesencephalon, which is faintly subdivided into two mesomeres. In accordance with BARTELMEZ (1923) the rhombencephalon is subdivided into five rhombomeres which are called RhA_1, rh_3, rh_4, rh_5, and *RhC*. The RhA_1 and *RhC* are larger than the others. The spinal tube displays twelve myelomeres in the somitic region. As far as can be seen, the tube is unsegmented between the caudalmost somite and the posterior neuropore. Thirteen distinct somites are developed. The rostral one is identical with the rostral one of the preceding stage (VAAGE and HØIVIK, 1969). Coal label experiments show that the pros-mesencephalic and the rhombo-spinal boundaries are identical with the arch-deuteroencephalic and the rhombo-spinal boundaries identified at stages *HH–9* and *HH–10*, respectively. Lateral to the rh_5 and the *RhC*, the ectoderm is depressed compared with the surroundings. The depressed area is the rudiment of the otic placode. Rostral to the prospective placode a condensation extends between the rh_4 and the lateral wall of the head. This condensation seems to be identical with the preotic neural crest of HIS (1868) and LILLIE (1919) and others. A lesser developed condensation occurs between the otic placode and the rostral somite and is identical with the rostralmost portion of the postotic neural crest of LILLE (1919). The rostral portion of the pros-rhombencephalic neural crest has disappeared, while the caudal portion is still located lateral to the RhA_1. No neural crest anlage is descernible lateral to the rh_3.

Stage HH–12. The prosencephalon is bent further ventrally (Figs. 8, 9). Thereby, the mesencephalon occupies a more cranial position compared with the preceding stage. The bulges of the neural tube are still more conspicuous than at stage *HH–11*. The RhA_1 is faintly subdivided into a rostral ($Is\text{–}rh_1$) and a caudal (rh_2) rhombomere (Fig. 9). Sixteen distinct somites are seen, and the spinal cord is subdivided into myelomeres in the somitic region. The otic pit is circumscribed. Lateral to the RhA_1, rh_4 and the *RhC* small condensations of cells form the pros-rhombencephalic, the preotic and the postotic portions of the neural crest, representing the rudiments of the semilunar, the facial-acoustic and the petro-nodose ganglions, respectively. The latter rudiment is somewhat irregular, being unsegmented in some specimens and segmented in others. The posterior neuropore is closed, and the spinal cord terminates in the primitive knot. The

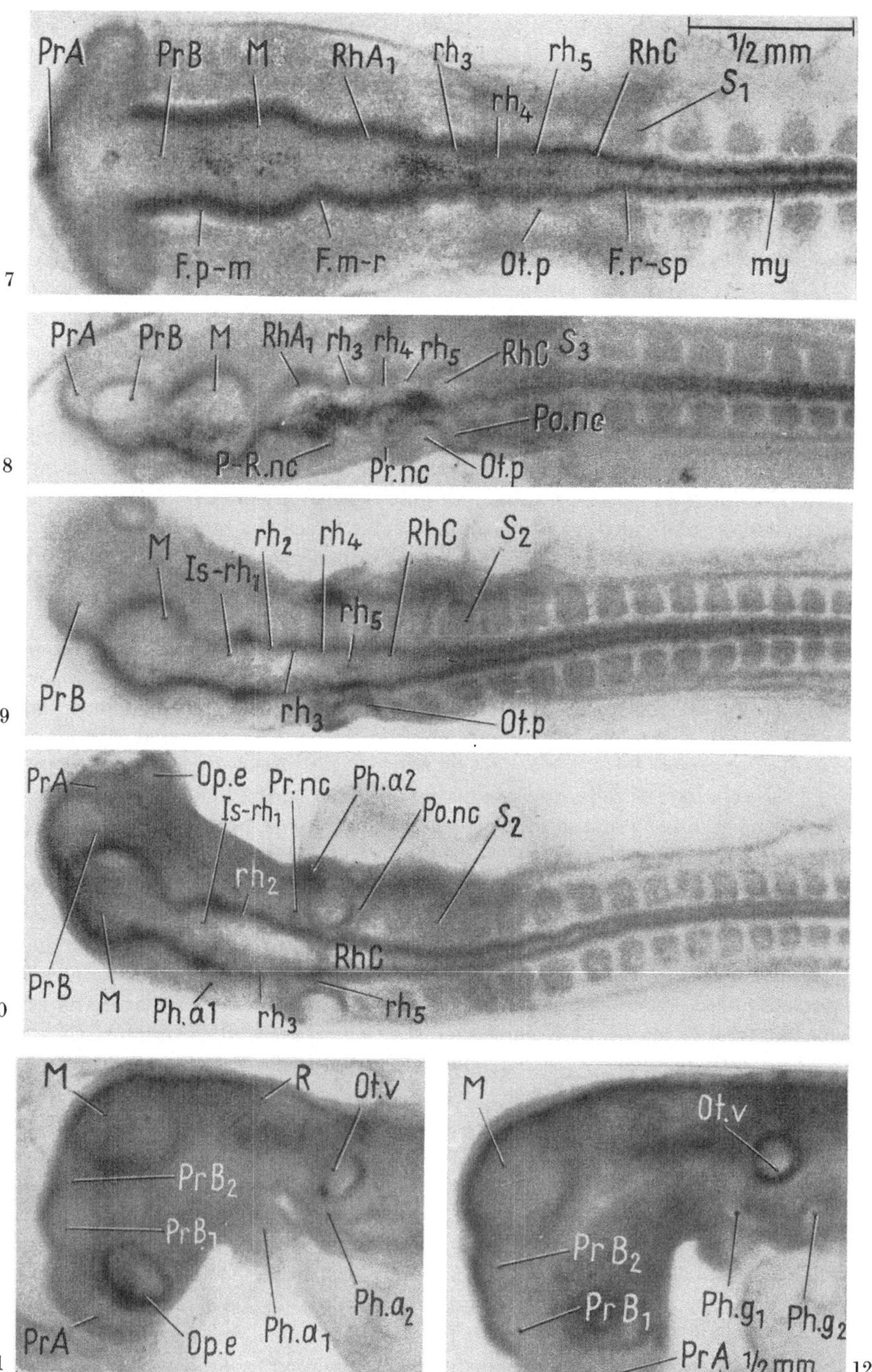

Figs. 7—12. Dorsal (Figs. 7—10) and lateral (Figs. 11, 12) views of the cerebral tube and the rostralmost portion of the spinal tube of living chick embryos at stages *HH–11*, *HH–12* (Figs. 8, 9), *HH–13* (Figs. 10, 11), *HH–14*

spinal cord interposed between the caudal pair of somites and the primitive knot is unsegmented.

Stage HH–13. The embryo bends left in the head region (Fig. 10), exposing the prosencephalon in lateral view, ventral to the mesencephalon (Figs. 10 and 11). The prosencephalon is subdivided into two (*PrA*, *PrB*, Fig. 10) or three prosomeres (*PrA*, PrB_1 and PrB_2, Fig. 11). The subdivision of the RhA_1 into a rostral (Is–rh_1) and a caudal (rh_2) rhombomere is more conspicuous. The rhombomeres rh_2 and rh_3 are broader than those immediately rostral and caudal to them. In a dorsal view (Fig. 10) the lateral wall of the rhombencephalon thereby forms a rhombe. The prospective semilunar ganglion is located lateral to the rh_2. The otic pit is circumscribed. Cell condensations rostral and caudal to the pit indicate rudiments of the cranial nerves and their ganglions. In the branchial region the mandibular and hyoid arches are distinguishable separated by the first pharyngeal groove. Nineteen somites are distinguishable (fourteen of which are visible on Fig. 10) and the spinal cord is divided into myelomeres in the somitic region. Caudal to the somites, no trace of myelomeres can be seen.

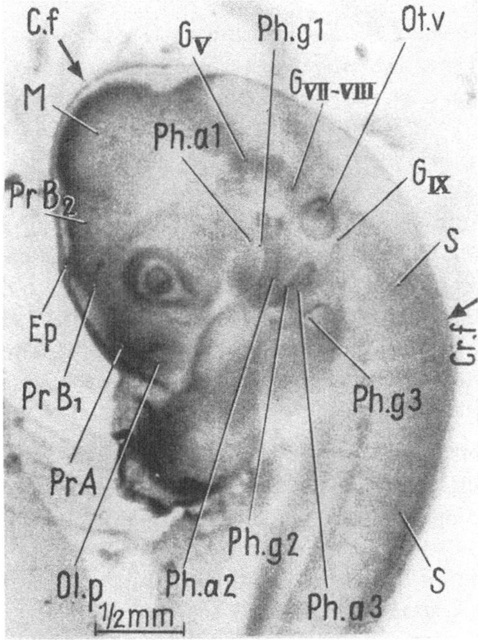

Fig. 13. Lateral view of a living chick embryo at stage *HH–16*

Stages HH–14 to HH–16. The prosencephalon is now definitely subdivided into three large bulges called *PrA*, PrB_1 and PrB_2 (Figs. 12 and 13). The olfactory pit is discernible on the ventral wall of the *PrA*. The segmental pattern of the rhombencephalon appears unchanged, but a lateral rotation of that particular region makes an accurate examination difficult. Three pharyngeal arches and grooves are present in the branchial region.

The occipital myelomeres have undergone reduction during this developmental period, but myelomeres are present further caudally in the somitic region of the tube. The spinal tube is unsegmented between the caudal somite and the end bud. The primitive knot is differentiated into the end bud and the tail bud in accordance with the pattern described by HOLMDAHL (1925).

Discussion

A. The Developmental Pattern of the Somites

At stage *HH–8–9*, the first intersomitic cleft (*I.s.c 1*), separating the two rostral somites, is located ventrolaterally to the first metameric segment in the spinal plate. From this stage onwards the rostral somite persists as the rostralmost for a considerable period (at least until *HH–14*). In the same period new somites are formed caudally in accordance with the general embryological pattern of differentiation, as shown previously by several authors (PATTERSON, 1907; HUBBART, 1908; WILLIAMS, 1910; LILLIE, 1919; JAGER, 1926; BEER and BERRINGTON, 1934; ROMANOFF, 1960). Experimental investigations have proved that the first intersomitic cleft persists from stage *HH–7* until stage *HH–14* or even longer (PATTERSON, 1907; HUBBART, 1908; VAAGE and HØIVIK, 1969). Therefore, this cleft may be used as a point of reference during the early developmental period. The somite situated rostral to the cleft is rudimentary (incomplete), while those caudal to the cleft are complete.

The first rudimentary somite is situated close to the neural tube. It appears at about *HH–7* and disappears later than stage *HH–14* (PATTERSON, 1907; HUBBART, 1908; JAGER, 1926; HINSCH and HAMILTON, 1956; VAAGE and HØIVIK, 1969). This somite and the first intersomitic cleft occur simultaneously, and both structures may therefore be used as landmarks during the early development of the central nervous system. (Concerning head somites, see GEGENBAUR, 1887; ZIEGLER, 1907 and GOODRICH, 1918.)

B. The Occurrence of the Neuromeres

The segmentation of the neural tube observed in living chick embryos definitely proves that the neural tube has a variable number of bulges during early ontogenesis. The morphology of the bulges (as seen in dissected and partly-sectioned embryos and in sections) leaves little doubt that the bulges are identical with the neuromeres described by ORR (1887). The study of the ventricular contours corroborates the findings of MALPIGHI (1673), v. BAER (1828), KÖLLIKER (1861), HIS (1868), MIHALKOVICS (1877), HILL (1900), WEBER (1900), v. KUPFFER (1906), HAMBURGER and HAMILTON (1951), BERGQUIST and KÄLLÉN (1954), BERGQUIST (1956, 1962), MENKENS (1967) and others. Since the neuromeres are observed in living, dissected and sectioned specimens it seems evident that they are intravitally existing structures and not artefacts.

However, divergent opinions prevail concerning the morphogenetic pattern of the neuromeres in both the cerebral and the spinal divisions of the neural tube.

a) The Encephalomeres

The demonstration *in vivo* of the two primary encephalomeres called archencephalon and deuteroencephalon, confirms the findings of GOETTE (1875), FOSTER and BALFOUR (1876) and v. KUPFFER (1906) among others. This observation is in accordance with the statement of MINOT (1892). v. KUPFFER (1906, p. 249) anticipated a similar developmental pattern when he stated ,,Der vordere kugelige Abschnitt der Vögel differenziert sich in die Augenblasen und das bleibende Vorderhirn, der hintere begreift das Mittel- und Rautenhirn zusammen".

Some authors [*e.g.*, BERGQUIST (1932, 1952a, b); KÄLLÉN (1954, 1965); BERGQUIST and KÄLLÉN (1954, 1955) and WEDIN (1954)] have rejected the morphological division of the cerebral rudiment into an archencephalic and a deuteroencephalic portion. As far as I am aware, these authors have not examined the two vesicle stage.

The conception of the archencephalon and the deuteroencephalon has assumed quite another significance through the experimental investigations carried out by NIEUWKOOP *et al.* (1952, 1955), EYAL-GILADI (1954), HARA (1961) and SAXÉN and TOIVONEN (1962) among others. Their experiments have shown that the differentiation of the neural anlage is governed by two different induction systems in the roof of archenteron (see below). The one system (the activating system) induces the formation of prosencephalic structures. The other system (the transforming system) causes the formation of mesencephalic, rhombencephalic and/or spinal cord elements (HARA, 1961; NIEUWKOOP, 1965). Both in amphibians (SALA, 1955) and birds (HARA, 1961) the prechordal mesoderm induces prosencephalic structures exclusively. In experiments where the presumptive notochordal mesoderm is used as inductor, mesencephalic and/or more caudal elements are formed. It seems likely, therefore, that the archencephalon and the deuteroencephalon in birds are early morphological expressions of the activating and the transforming principles, respectively. In my opinion, these induction experiments confirm observations showing that the mes- and rhombencephalon are epichordal brain regions, while the prosencephalon is prechordal.

Shortly after its formation, the deuteroencephalon subdivides into a lesser rostral encephalomere, the mesencephalon, and a larger caudal one, the rhombencephalon. The cerebral tube is then subdivided into three encephalomeres (vesiculae cerebrales), *viz.*, the pros-, mes- and the rhombencephalon. Concerning the names used earlier, see MIHALKOVICS (1877). This morphogenetic pattern of the mesencephalon in birds is in accordance with the observations of FOSTER and BALFOUR (1876) and v. KUPFFER (1906).

However, authors have interpreted the encephalomeres at the early developmental stages in a different way. If figure 1 of BERGQUIST (1956) is compared with the findings of HILL (1900) and PATTEN (1958), it seems obvious that the bulge designated as neur. d (VI) by BERGQUIST is identical with the bulge called mesencephalon by HILL and PATTEN. In view of my observations the encephalomere called metencephalon by BERANECK (1887) and BERGQUIST (1956) at the 10—12 somite stage is identical with the bulge named mesencephalon by HIS (1868), MIHALKOVICS (1877), HILL (1900) and PATTEN (1958) at the same developmental stage.

Each encephalomere is subsequently subdivided and transformed and some of the bulges can be observed *in vivo*. The subdivision of the rhombencephalon starts at *HH–9+* simultaneously with the subdivision of the deuteroencephalon into the mes- and rhombencephalon. The latter consists of three (*RhA*, *RhB* and *RhC*) or four (RhA_1, rh_3, *RhB* and *RhC*) bulges from the very beginning of subdivision. The prosencephalon starts to divide prior to stage *HH–11*. In other words, the rhombencephalon begins to subdivide before the prosencephalon. The subsequent further segmentation of the two regions takes place at different stages and proceeds at different rates (see chapter V).

Bergquist, Källén and co-workers found formation and subsequent disappearance of three different waves of segmentation, which they called proneuromeres, neuromeres and postneuromeres. Hence the encephalomeres disappear three times during the ontogenesis of the brain, and each time in a rostrocaudal direction (Bergquist and Källén, 1954, 1955). According to my observations, however, no reduction of the neuromeres takes place during interneuromeric phases one and two of Bergquist and Källén (1954, 1955). But the neuromeres subdivide and are transformed during the early developmental stages. This transformation is discussed further in chapter V.

b) The Myelomeres

The rostral myelomere is situated between the first rudimentary somite and the first complete one, that is, the centre of the myelomere lies opposite the intersomitic space. The succeeding myelomeres are formed simultaneously with the somites as observed by Minot (1892, p. 605) among others, who stated that "the neuromeres are readily seen to correspond exactly as do later the nerves, with the number of segments". Consequently the myelomeres are formed in one rostrocaudal wave of segmentation bulges. These observations have been confirmed repeatedly in a variety of vertebrates (Minot, 1892; Neal, 1898; 1918; Johnston, 1916; Streeter, 1933).

However, Bergquist, Källén and their collaborators advocate a different view concerning the myelomeres (Källén and Lindskog, 1953; Wedin, 1954; Bergquist and Källén, 1954, 1955; Bergquist, 1956, 1962; Källén, 1956a—c, 1965). They maintain that the myelomeres are formed in two waves of bulges, called by them the proneuromery and the neuromery waves (Källén and Lindskog, 1953). The two segmentation waves in the spinal cord may be called the promyelomery and the myelomery waves, respectively, adapting the nomenclature introduced by McClure (1890). In the description of the promyelomeres and the myelomeres their relationship to the somites is not mentioned by Källén and Lindskog (1953). No somites are described, nor outlined in their figures. In chick embryos the promyelomeres and the myelomeres are said to occur between *HH–8*, *HH–9* and *HH–10*, *HH–13*, respectively. From *HH–13* onwards, the spinal cord is not subdivided into myelomeres at all. In my material, the rostral myelomeres occur rostrally in the spinal cord at *HH–9*, and they occur more caudally in the spinal cord at least as late as stage *HH–16*. No intervening disappearance of the myelomeres is observed. According to Lillie (1919) the segmentation continues until *HH–28–29* when the tail region has become segmented.

At the stages *HH–9* and *HH–13*, only the rostral part of the spinal cord is formed. It terminates in the primitive knot (*Hensens knot*) and later on in the end bud (HOLMDAHL, 1925). The promyelomeres and myelomeres of BERGQUIST and KÄLLÉN thus occur only in the rostral half of the spinal cord, since the caudal half is formed later in ontogenesis. My observations in the chick show that the myelomeres are formed by one rostrocaudally moving segmentation wave occurring concomitantly with the formation of the somites. The spinal cord caudal to the somites is not segmented as far as I am aware. In the observations on living specimens I find that myelomeres occur for a longer developmental period than dexcribed by BERGQUIST and KÄLLÉN (1953a, 1954, 1955). The rostral myelomeres are distinguishable from *HH–9* until *HH–13–14*, *i.e.* during the same developmental period as the promyelomery and the myelomery waves. The more caudal myelomeres have a developmental period similar to the rostral ones, since the disappearance of the myelomeres takes place rostrocaudally (see below), and not caudorostrally as described by KÄLLÉN and LINDSKOG (1953). These authors state that the caudal myelomeres have the shortest duration, being the last formed and the first ones to disappear. No intermyelomery phases I and II (of BERGQUIST and KÄLLÉN) are observed by me in chick.

Conclusion

The neural tube in living chick embryos is subdivided into bulges, in the literature called neuromeres.

The cerebral anlage differentiates in a caudorostral and the spinal cord in a rostrocaudal direction, concomitantly with the somites.

IV. Observations on Fixed Embryos Dissected in the Fixative, and on Wax Plate Reconstructions

The aforementioned observations on early stages of living specimens give no information concerning the extent of the superficial furrows interposed between the neuromeres. The external surface of the neural tube and the dorsoventral extent of its bulges and furrows can be studied in fixed embryos and wax plate reconstructions. In the following, the morphological development of stages *HH–8–9* to *HH–16* is described in some detail.

Results

Stages HH–8–9 to HH–9+. A superficial furrow (*F.r–sp*) medial to the rostral somite subdivides the neural tube into a cerebral and a spinal portion (Fig. 14). This is easily seen when the rostral somite is dissected away. The cerebral tube, which is broad rostrally and narrow caudally, clearly shows the archencephalon and the deuteroencephalon observed *in vivo*. The optic evagination occupies the whole of the lateral wall of the archencephalon. The deuteroencephalon is broadest rostrally and continues into the spinal cord without a distinct transition zone.

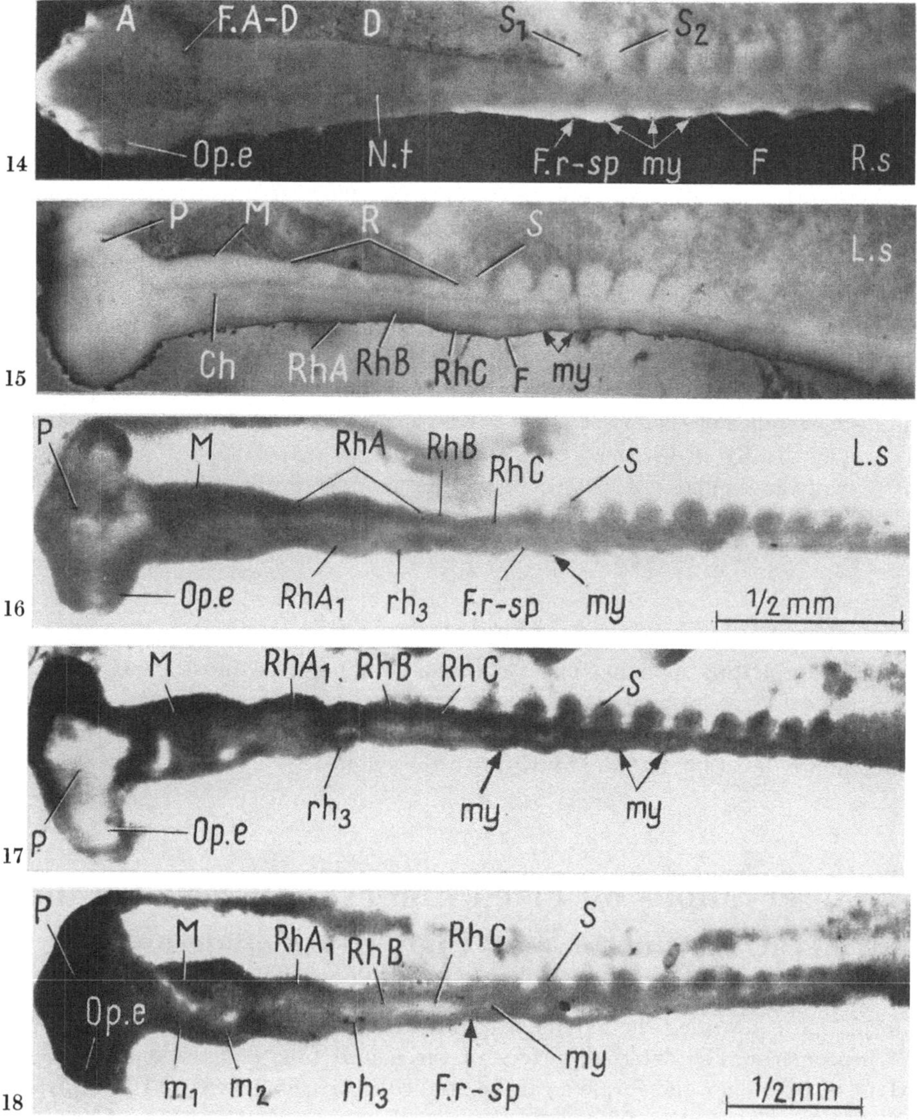

Figs. 14—18. Ventral views of the neural tube of chick embryos at the stages *HH–8–9*, *HH–9*, *HH–9–10*, *HH–10* (Figs. 17, 18). All somites are dissected away on the right side (*R.s*)

The lateral wall of the neural tube is subdivided into segments which according to Locy (1895) and Hill (1900) can be traced (stage by stage) from identical segments of preceding stages. The segments are especially distinctly developed in the prospective spinal cord which consists of several myelomeres (my Fig. 14). The myelomeres are separated from each other by superficial furrows corresponding to the middle of each somite. Thus there is only one myelomere interposed between two successive somites.

The arch-deuteroencephalic as well as the rhombo-spinal boundary are orientated at right angles to the longitudinal axis of the neural tube. At stage *HH–9* (Fig. 15), the archencephalon (from this stage called the prosencephalon, *P*) is unsegmented dorsally and laterally, but ventrally it is subdivided into three segments derived from the neural plate segments, as described by LOCY (1895) and HILL (1900). The optic evagination is situated more rostrally than at the preceding stage. The deuteroencephalon is subdivided into a lesser rostral encephalomere, the mesencephalon, and a larger caudal one, the rhombencephalon. The latter consists of three rhombomeres (*RhA*, *RhB* and *RhC*), obviously identical with those observed *in vivo* at stage *HH–10*. The *RhB* is the narrowest one. At stage *HH–9–10* (Fig. 16), the *RhA* is subdivided into a rostral (RhA_1) and a caudal (rh_3) rhombomere. The myelomeres occur at all three stages.

Stage HH–10. The prosencephalon has grown broader rostrally (Figs. 17 and 18) compared with the preceding stages. The caudal half is narrow. The bulges in the mes- und the rhombencephalon are more prominent than at stage *HH–9–10*. The *RhB* is subdivided into the rh_4 and rh_5 at stage *HH–10–11*. The myelomeres are distinct (*HH–10*: Figs. 17 and 18). New myelomeres have been added caudally. The notochord terminates at the boundary between the pros- and the mesencephalon (Fig. 15).

Stage HH–11. The prosencephalon is dorsally subdivided into two bulges (Figs. 19, 21) by the prospective transverse septum. The furrow ($F.t\text{–}d_1$, Fig. 25) runs just caudal to the optic evagination which belongs to the rostral prosomere (*PrA*). This bulge bends ventrally compared with the mes- and rhombencephalon. The caudal prosomere, the *PrB*, is more narrow than the rostral one. The mesencephalon is still subdivided into two mesomeres, and it is compressed rostrocaudally compared with the rhombencephalon. The latter consists of five bulges (RhA_1, rh_3, rh_4, rh_5, *RhC*). The rh_4 is situated somewhat more ventrally than the rhombomeres rostral and caudal to it. The rostrocaudal length of the rh_4 is greater ventrally than dorsally. The rh_5 on the other hand, is shorter ventrally than dorsolaterally. The rh_4 is, therefore, easy to discover in the lateral and ventral view, and it is a marker in the rhombencephalon. The *RhC* is dorsally subdivided into rh_6 and rh_7.

Myelomeres are present in the entire somitic region of the spinal cord.

Stage HH–12. The cerebral tube bends in the mesencephalic region indicating the cranial flexure (Fig. 22). In lateral view, the mesencephalon occupies the dorsalmost position compared with the pros- and the rhombencephalon. The wall of the *PrA* bulges dorsolateralwards indicating the prospective spherencephalon of NEUMAYER (1899). The mesencephalon has achieved a more spherical form, and no subdivision is visible on the surface.

The rhombencephalon has undergone no change from stage *HH–11*. Owing to the cranial flexure the boundary ($F.t\text{–}d_1$, Fig. 25) between the *PrA* and the *PrB* extends approximately parallel to the longitudinal axis of the rhombencephalon. The boundaries (Fig. 25) between the major brain regions are unaltered from the preceding stages.

Stage HH–13. The cranial flexure has increased (Figs. 22 and 23). The *PrA* has enlarged dorsally and laterally. The *PrB* is elongated and subdivided into two prosomeres (PrB_1 and PrB_2) which are identical with the two prosomeres

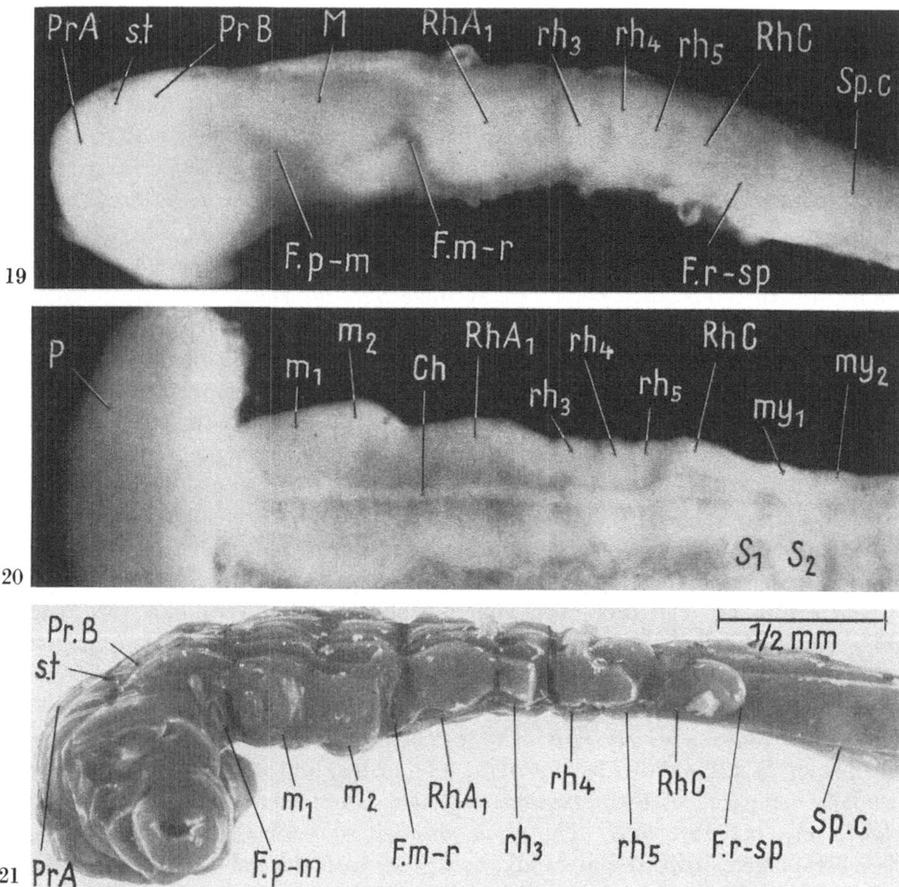

Figs. 19 and 20. Lateral (Fig. 19) and ventral (Fig. 20) views of the neural tube of chick embryo at stage *HH–11*. The somites are dissected away on the left side

Fig. 21. Wax-plate reconstruction of the neural tube of chick embryo at stage *HH–11*

observed in the same region *in vivo* from stage *HH–13* on (Fig. 11). The mesencephalon bulges out dorsally. In the roof of the rhombencephalon the roof plate has become broad and consists of a flattened layer of cells. The roof plate enlarges greatly in succeeding stages. Rostrally in the spinal cord the furrows between the myelomeres are reduced. The boundaries between the major subdivisions of the brain anlage are distinguishable but less well-defined than at preceding stages.

Stage HH–16. The neural tube has expanded considerably (Fig. 24). The dorsal furrow between *PrA* and *PrB* has grown deeper and forms the prospective transverse septum, but its lateral extension into the $F.t\text{–}d_1$, has disappeared. The lateral wall of the *PrA* bulges out dorsally to form a separate bulge called pr_3. The optic stalk terminates ventrolaterally within this bulge. A furrow ($F.t\text{–}d_2$) starts dorsally within the transverse septum, continues ventralwards rostral to the pr_3 and terminates ventrally, rostral to the optic nerve. This sulcus is pre-

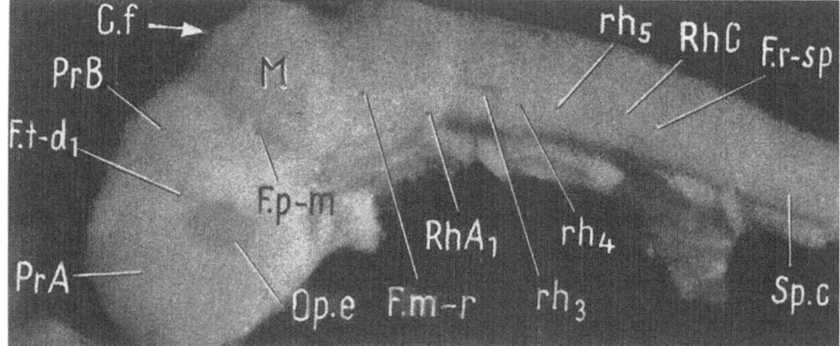

22

23

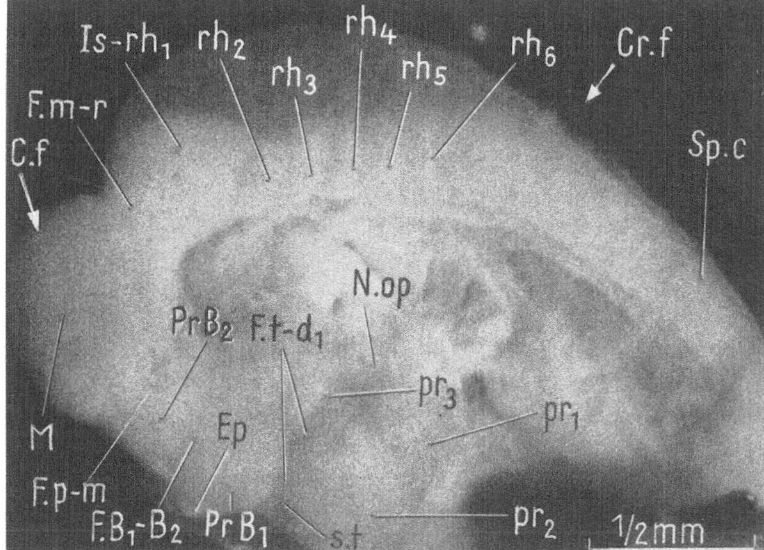

24

Figs. 22—24. Lateral views of the neural tube of chick embryos at stages *HH-12*, *HH-13* and *HH-16*. The somites are dissected away on the left side

sumably identical with the tel-diencephalic sulcus of KAMON (1906) and MEEK (1907). A shallow furrow below the nasal placode subdivides the rostral portion of the *PrA* into a rostroventral pr_1 and a dorsal pr_2. The prospective spherencephalon is located dorsally within the pr_2. The spherencephalon expands dorsolaterally on both sides of the pr_2 in the succeeding stages. The furrow between the PrB_1 and PrB_2 extends dorsoventrally to terminate in the region of the posterior tubercle. The rudiment of the pineal gland grows out from the top of the PrB_1. The mesencephalon consists of one spherical bulge which, at this stage, occupies the rostralmost position of the neural tube. The pros-mesencephalic and the mes-rhombencephalic boundaries are well defined throughout all developmental stages investigated, as shown in Fig. 25. The rhombencephalon is subdivided into six distinct rhombomeres called $Is\text{-}rh_1$, rh_2–rh_6. The rhombo-spinal boundary has faded, and the neural tube medial to the rostralmost somites has expanded. In the somitic region of the spinal cord a new bordering line occurs at later stages. The spinal cord located rostral to this boundary belongs to the myelencephalon, while that caudal to the line differentiates into the definitive spinal cord (Fig. 25).

Discussion

The Relationship between the Various Intracerebral Boundaries and the Neuromeres

There can be no doubt that the bulges observed in dissected embryos are identical with those observed in living specimens. Similar bulges are also observed in specimens dissected in amniotic fluid. Therefore, the bulges do not represent artefacts caused by the fixation agent.

The subdivision of the brain is based on the segmentation of the primitive neural tube.

a) The Tel-Diencephalic Boundary

The prosencephalon is initially subdivided into *PrA* and *PrB* by a dorsoventral furrow ($F.t\text{–}d_1$, Fig. 25). This boundary disappears during the neurogenesis, and another one develops within the lateral wall of the *PrA* ($F.t\text{–}d_2$, Fig. 25), separating the pr_{1-2} from the rest of the *PrA*. Both these furrows have a corresponding ventricular ridge (see next chapter) and they represent boundary furrows. However, in the literature, the two furrows are both called the tel-diencephalic boundary. In the opinion of some authors, the tel-diencephalic boundary starts dorsally at the transverse septum and extends along the boundary between the two primary prosomeres, terminating ventrally between the infundibular anlage and the mammillary recess. The optic anlage and the optic stalk belong to the rostral prosomere as advocated by WEBER (1900) and KAMON (1906) among others. My observations favour the view that this boundary is identical with the first inter-prosomeric boundary. The furrow is merely a transitory one and disappears later on. In the opinion of other authors, the tel-diencephalic boundary extends along the lateral wall of the spherencephalon. The boundary is marked dorsally by the transverse septum and extends ventrally to terminate rostral to the optic evagination and the optic stalk (WATERS, 1892; HILL, 1900; MEEK, 1907; BERGQUIST and KÄLLÉN, 1954) or in the optic recess (v. KUPFFER, 1906). As consequence of this view, the optic anlage is restricted to the diencephalon. According to my observations, this boundary is identical with the furrow between pr_{1-2} and the rest of the *PrA* ($F.t$–d_2, Fig. 25). The question now arises: Which of these boundaries represents the tel-diencephalic border? This is, in my opinion, a question of conventional description, since both boundaries disappear during the ontogenesis. (For further consideration, see General Discussion.)

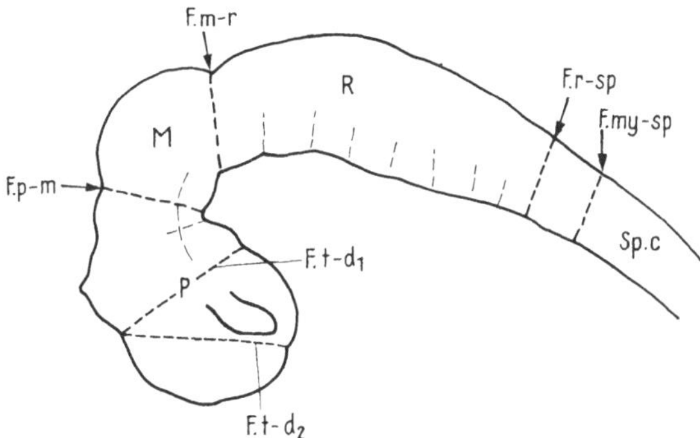

Fig. 25. Diagrammatic lateral view of the neural tube in chick showing the location of the bordering furrows

b) The Di-Mesencephalic Boundary

This boundary is identical with the pros-mesencephalic border. In 1898, Neal established that the boundary runs just caudal to the posterior commissure as confirmed later by v. Kupffer (1906) and Palmgren (1921) and supported by my results (see below). My observations in living and dissected chick specimens show that the arch-deuteroencephalic boundary is identical with the pros-mesencephalic and the di-mesencephalic boundaries.

c) The Mes-Rhombencephalic Boundary

The mes-rhombencephalic boundary is distinguishable from the three vesicle stage onwards and easily determined in subsequent developmental stages. The caudal mesomere (see next chapters) undergoes reduction and is transformed into a boundary zone between the mesencephalon and the rhombencephalon. The boundary is a cellfree mantle and a ventricular sulcus. The isthmic fovea develops mid-ventrally just caudal to the m_2 segment. The floor plate, the raphe medullae, the distribution of the glycogen ventrally in the neural tube and the non-proliferating cell column (*NPC*) all extend rostralwards to the level of fovea isthmi (see next chapters). The notochord continues farther rostralwards to terminate in the prechordal plate ventral to the caudal part of the diencephalon. The boundary segment (m_2) is interposed between the oculomotor and the trochlear nuclei, situated within the mesencephalon and the rhombencephalon, respectively. All the same, several authors (Locy, 1895; Hill, 1900; Meek, 1907; Hugosson, 1957), using different criteria, have placed the mes-rhombencephalic boundary in birds caudal to the trochlear nucleus. Thus Hugosson says: "On a level with the nuclear anlage there is no septum medullae and the anlage seems, therefore, to have its origin within the mesencephalon" (l.c., p. 75—76), and later in the ontogenesis — "the nucleus appears, moreover, to have wandered somewhat rostralward within the mesencephalon" (l.c., p. 76).

According to Kingsbury (1922) and the observations of the present author no septum medullae is developed until stage *HH–24*. Consequently the septum cannot be used as a criterion at earlier stages. Locy (1895), Hill (1900) and Meek (1907), did not investigate the relationship between the mes-rhombencephalic boundary and the differentiating trochlear nucleus in closely graded developmental stages. My observations in dissected and sectioned specimens show that the trochlear nucleus both develops and remains within the rhombencephalon (see next chapters).

d) The Rhombo-Spinal Boundary

It is appropriate in amniota to distinguish between a primitive rhombo-spinal and a secondary myelo-spinal boundary.

α) The Rhombo-Spinal Boundary. The embryological rhombo-spinal boundary separates the brain rudiment and the spinal cord anlage. This boundary (Fig. 25) is, at early ontogenetic stages, located medial to the rostral somite (the first rudimentary somite). The boundary, therefore, is distinguishable from the two somite stage onwards when the rhombencephalon „sich verengend bei dem ersten Urwirbelpaar in das Rückenmarksrohr (*spn*) übergeht" (Mihalkovics, 1877, p. 22). A similar "embryological" definition of the rhombo-spinal boundary is used by Malpighi (1673), v. Baer (1828), Minot (1892), Locy (1895), Hill (1900), and others. In chick this boudary separates *RhC* and the first myelomere (my_1) until stage *HH–13* when the furrow disappears.

β) The Myelo-Spinal Boundary. From stage *HH–13* on, the rostral myelomeres merge with the caudal rhombomere (see chapter IV) to form a large "rhombomere" called "rh_7", which constitutes a part of the myelencephalon. The rostral somites are situated laterally to this new bulge. Several myelomeres merge through this process with the myelencephalon. Later on, a new boundary appears between the myelencephalon and the spinal cord, called the myelo-spinal boundary. However, some embryologists (Minot, 1892; Meek, 1907; Lillie, 1919 and Holmdahl, 1928) have used the term rhombo-spinal for the myelo-spinal boundary. Thus Lillie (1919, p. 126) states: "We may distinguish a cephalic portion (brain or encephalon) and a trunk portion (spinal cord or myelon) of the neural tube, the boundary lies between the fourth and fifth somites, for the first four somites enter into composition of the head."

In my opinion it is misleading to call the myelo-spinal boundary the rhombospinal. From an embryological point of view, it seems appropriate to reserve the designation rhombo-spinal boundary for the embryological boundary between the primitive rhombencephalon and the spinal tube, for the following reasons:

αα) The myelomeres are serially homologous structures, which are formed rostrocaudally and concomitantly with the formation of the somites. They have segment nerves (Meek, 1907). The myelomeres, the spinal nerves and the (metameric) somites are presumably an indication of the fundamental trunk metamerism.

$\beta\beta$) The formation of the rhombomeres and myelomeres is induced from two different zones in the roof of the archenteron (LEHMANN, 1945). The myelomeres have a developmental pattern different from the rhombomeres, and they are not homologous structures according to NEAL (1918).

The rhombomeres occur independently of the head metamerism and the branchiomerism (see General Discussion p. 75).

$\gamma\gamma$) The merging of the myelomeres with the caudal rhombomere may be looked upon as an evolution within the vertebrate series parallel to the evolution of the spinal accessory and the hypoglossal nerves. Consequently different numbers of myelomeres are merging with the rhombencephalon: none in the anamniota, 4—7 in reptiles and birds, and 3—4 in mammals (see next chapter).

$\delta\delta$) An identical rhombo-spinal boundary (situated medial to the rostralmost somite, *i.e.* the first, incomplete somite) has been observed and described by embryologists since MARCELLO MALPIGHI observed it in birds in 1673 (cited from TIEDEMANN, 1816).

Conclusion

At early stages the tel-diencephalic boundary is located caudal to the infundibular region ($F.t\text{–}d_1$) but is later replaced by a border located rostral to the optic nerve ($F.t\text{–}d_2$).

The pros-mesencephalic boundary (= di-mesencephalic boundary) is situated just caudal to the posterior commissure.

The mes-rhombencephalic boundary extends between the oculomotor and the trochlear nuclei.

The rhombo-spinal boundary is located medial to the rostralmost metameric somite.

V. Observations on Fixed and Partly-Sectioned Embryos

The material described in the preceding chapters have revealed the external morphology of the neural tube, but have not contributed materially to shedding light on its ventricular configuration. Information on this subject is obtained by using the procedure of partial dissection following embedding in paraffin, as appears from the following description of stages *HH–10* to *HH–27*.

Results

Stage HH–10. A number of distinct ventricular sulci are separated by dorsoventrally running ridges (Fig. 26). Some of the sulci and ridges evidently correspond to the superficial bulges and furrows observed in living and dissected specimens. In the ventral half of the cerebral tube several transversal sulci are discernible, gradually disappearing dorsolaterally. Some of the sulci may be identical with the primary neuromeres of HILL (1900). The prospective transverse septum divides the prosencephalon dorsally into two prosomeres (*PrA* and *PrB*). Two mesomeres (m_1 and m_2) and five rhombomeres (RhA_1, rh_3–rh_5, RhC) are distinguishable.

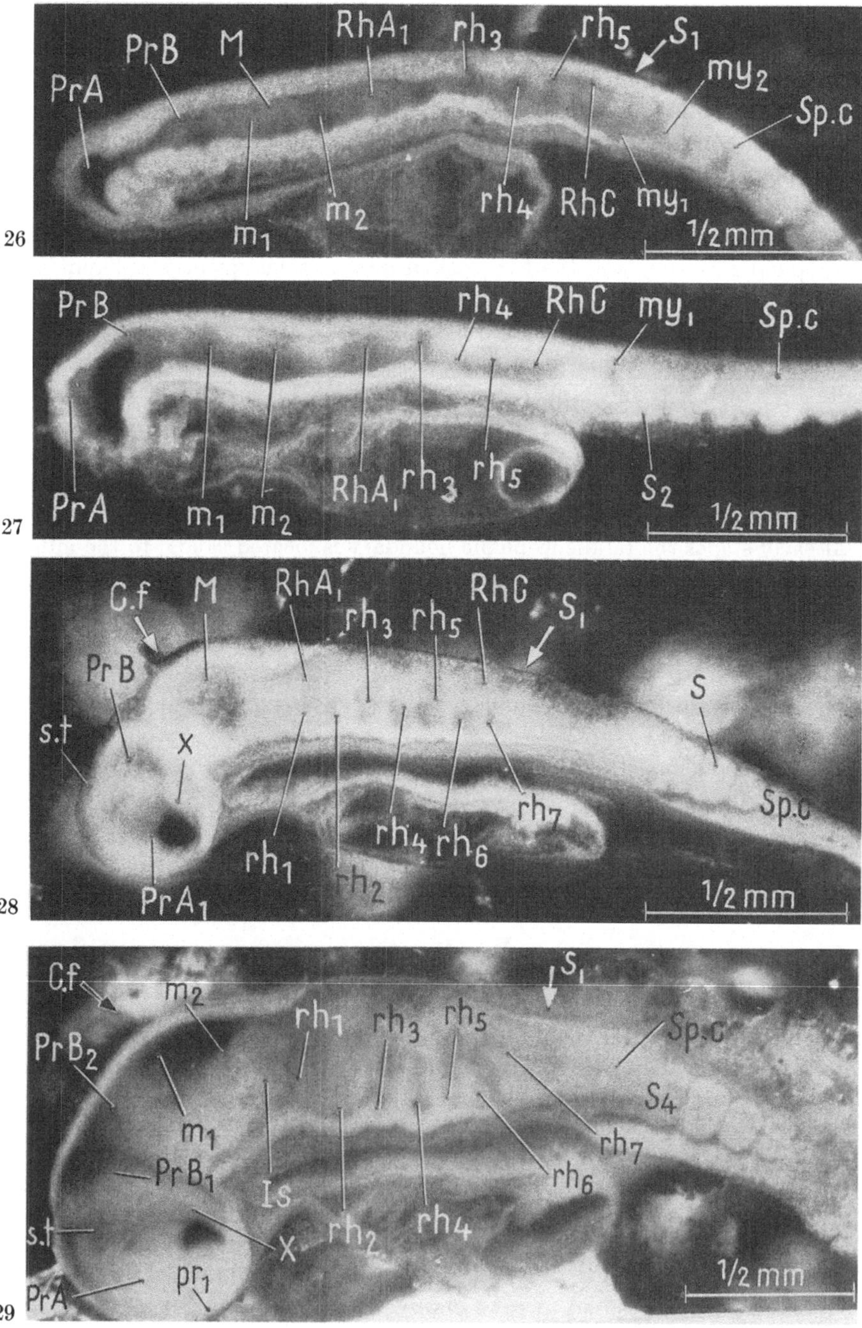

Figs. 26—29. The right ventricular surfaces of the cerebral tubes of chick embryos at stages *HH-10*, *HH-11*, *HH-12* and *HH-13*, exposed by removal of the left half of the tube

In the spinal cord a myelomere is located between two succeeding pairs of somites. Faintly developed ventricular ridges correspond to the bordering furrows between the prosencephalon, the mesencephalon, the rhombencephalon and the spinal cord.

Stage HH–11. In the floor of the cerebral tube (Fig. 27) shallow remnants of transversal sulci are seen. The sulci undergo reduction in subsequent stages (Figs. 28 and 29) and, according to ORR (1887), these particular sulci are of no morphological significance. The prosencephalon is bent ventrally, and the *PrA* is situated ventral to the *PrB*. Two distinct mesomeres are discernible.

Stage HH–12. The cranial flexure is indicated in the mesencephalic region (Fig. 28). The interprosomeric ridge between the *PrA* and *PrB* begins dorsally at the prospective transverse septum, runs laterally, and fades little by little ventrally. If lengthened ventrally, the ridge will terminate just rostral to the mammillary recess which consequently belongs to the *PrB*. Ventrally within the *PrA* a ridge (x, Fig. 28) transverses the neural tube just caudal to the optic evagination and continues caudalwards into the ventrolateral wall of the *PrB*. The ridge is probably identical with the "Basilarleiste" of HIS (1868) and separates the infundibular anlage (pr_4) from the central portion of the *PrA*. The PrA_1-sulcus begins dorsally in the rudiment of the spherencephalon and runs ventralwards into the optic evagination and the optic recess. Both the spherencephalon and the optic anlage are situated within the PrA_1. In the mesencephalon the m_1 has grown dorsally and laterally, while the caudal mesomere lags behind. The rostral portion of the RhA_1 is long and narrow and has no distinct ventricular sulcus. In the caudal portion, two sulci occur within two separate segments termed rh_1 and rh_2. No distinct ventricular ridge separates the Is from the rh_1. The rh_4 is longer and lies further ventrally than the rh_3 and rh_5. However, it narrows in a dorsolateral direction where the rh_5 is correspondingly longer. Thus the ventricular configuration of the rh_4 is in full accord with the external morphological features as revealed by dissection. Its characteristic morphology makes the rh_4 a conspicuous landmark from this stage onwards, both on the outer and the inner surfaces of the rhombencephalon. Ventrally the *RhC* is faintly subdivided into two rhombomeres (rh_6 and rh_7).

Stage HH–13. The cranial flexure is more conspicuous (Fig. 29). Rostral and dorsal to the optic evagination the tube bulges faintly, forming a shallow sulcus called the pr_1-sulcus. The *PrB* forms two bulges, a large rostral (PrB_1) and a small caudal one (PrB_2), as observed in living and dissected specimens. The observations confirm that the PrB_1-sulcus terminates in the mammillary recess situated rostrally to the posterior tubercle, while the PrB_2-sulcus, which is largest dorsally, terminates caudal to it. The ventral longitudinal ridge in the prosencephalon (x, Fig. 29) is reduced somewhat. Owing to the cranial flexure, the mesencephalon occupies a position relatively more rostral than earlier.

The rostral mesomere (m_1) has grown further dorsally, while the caudal one (m_2) is reduced to a narrow band-like sulcus. The bulges *PrB* and mesencephalon in the figures (Figs. 28 and 29) are identical with the mesencephalic and metencephalic bulges, respectively, described by AREY (1965, Fig. 545). The divergent interpretations of the prospective mesencephalon by various authors (BERANECK, 1887; BERGQUIST, 1956; AREY, 1965), may be the consequence of the relative shift in position of the bulges in the pros- and mesencephalon during this developmental period. The rostral rhombomeres (*Is*-rh_5) are distinct. The two sulci located ventrally within the *RhC* have grown, and they subdivide the particular rhombomere into the rh_6 and rh_7 homologous to those in mammals. They have been described by Bartelmez (1923) and Adelmann (1925). Identical rhombomeres are repeatedly described in birds (McClure, 1890; Zimmermann, 1891; Kamon, 1906; Neal, 1918). The ventricular ridges between the rostralmost myelomeres are less pronounced than in the preceding stages. Further caudally in the spinal cord, the myelomeres are well-defined.

Stage HH–14. The cranial flexure is more pronounced than at stage *HH–13* Fig. 30). A shallow sulcus (the prospective pr_3-sulcus) begins in the optic evagination and continues dorsalwards towards the transverse septum. The PrB_1 has become enlarged, especially dorsally. The roof of the rhombencephalon is dilated and membranous. The rostral myelomeres are less distinct and may be separated from each other only dorsally. The Rathke pouch is demarcated ventral to the pr_4. The notochord extends below the rhomb- and mesencephalon to terminate in the prechordal plate between Seessel's pocket and the mammillary recess.

Stage HH–15. The cranial flexure has increased still more. A second flexure (the cervical flexure) is indicated in the rostralmost portion of the spinal cord (Fig. 31). The PrA_1 is on the ventricular surface subdivided into a dorsal (pr_2) and a ventral (pr_3) portion by a ridge beginning dorsolaterally at the transverse septum and running laterally and ventrally. The ridge terminates ventromedially just rostral to the optic evagination, conforming with the pattern observed in dissected specimens. The ridge divides the PrA_1-sulcus in two. The dorsal division lies within the spherencephalon, while the ventral division extends from the optic recess lateralwards where it fades gradually. The two sulci are situated within the pr_2 and pr_3, respectively. The ridge interposed between them seems to correspond to the superficial furrow separating the pr_{1-2} from the pr_{3-4} observed in dissected specimens. In the ventral half of the PrB_1 two sulci extend dorsoventrally. Within the prosomeres pr_5 and pr_6, the rostral sulcus runs into the mammillary recess, while the caudal sulcus is located within the caudal portion of the PrB_1. The epiphysis grows out from the dorsal wall of the pr_6. The roof of m_1 has expanded further. In the rhombencephalon six ventricular sulci are separated by ridges. The rh_7 as well as the rostralmost myelomeres are reduced; the ridge between the rh_7 and the spinal cord has vanished, leaving a wide sulcus which superficially is marked by the bulge "rh_7" observed in dissected specimens. In the caudal half of the rhombencephalon and rostrally in the spinal cord, a longitudinal sulcus (*S.rh*) is discernible, presumably identical with the longitudinal sulcus described by Kamon (1906) and others.

Stage HH–17. The ventricular ridge between the pr_2 and pr_3 extends laterally and dorsally, merging with the transverse septum (Fig. 32). This ridge is the most conspicuous one in the prosencephalon and subdivides it into a rostral portion comprising the pr_1 and pr_2 and a caudal portion consisting of the pr_{3-8}. The sulcus within the pr_3 disappears gradually on the lateral wall. The pr_4 has grown larger and its sulcus runs further laterally and dorsally than earlier. The pr_5-sulcus merges with the mammillary recess. The PrB_2 is subdivided into pr_7 and pr_8. The prosomeric sulci are very distinctly developed ventrally and ventrolaterally and fade dorsally. The caudal mesomere encircles the neural tube, terminating in a shallow recess mid-dorsally. The *Is* and rh_1 are separated by a faintly developed

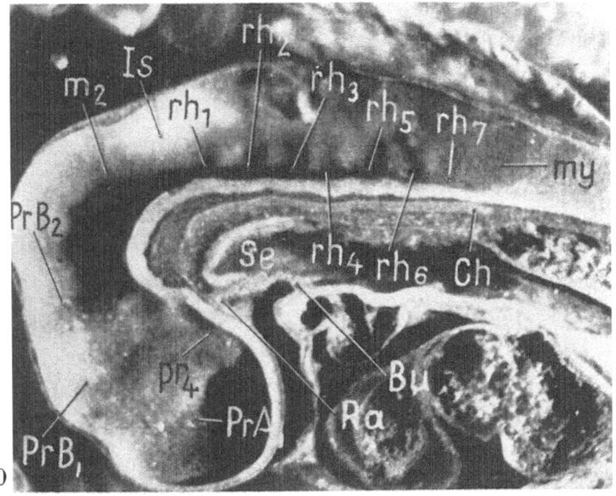

30

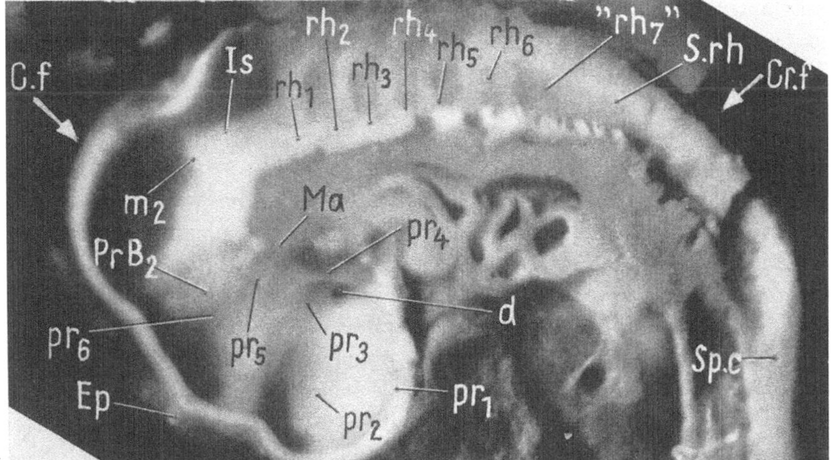

31

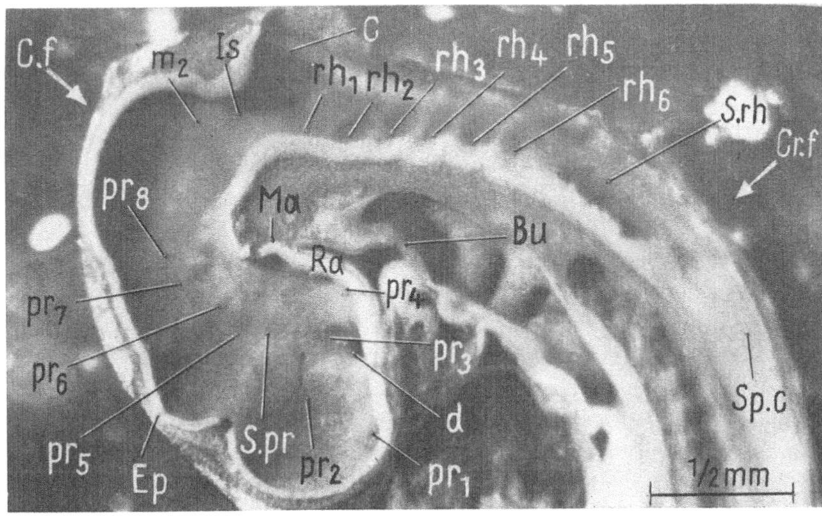

32

Figs. 30—32. The right ventricular surfaces of the cerebral tubes of chick embryos at stages *HH–14*, *HH–15* and *HH–17*, exposed by removal of the left half of the tube dorsoventral ridge, and they have a similar ventricular sulcus, as in succeeding rhombomeres. The *Is* is, however, narrow and constitutes the isthmus interposed between the mesencephalon and the rhombencephalon. The cerebellar rudiment occupies the dorsal portion of the rh_1. The rhombomeres are distinct. The rostralmost myelomeres are indistinguishable, while the more caudal ones are well marked. The rhombencephalic longitudinal sulcus (*S.rh*) discernible at the preceding stage, has grown farther caudalwards and rostralwards, where it terminates in the rh_1-rh_2. Another longitudinal sulcus (*S.pr*) in the lateral part of the pr_3 extends across the lateral wall of the prosomeres and into the rostral mesomere. No connection exists between the two longitudinal sulci as far as can be seen. The *Rathke pouch* has obtained its definite relationship to the pr_4 (the infundibulum) terminating just rostral to the mammillary recess.

Stage HH–20. The pr_1 is relatively reduced and the boundary between the pr_1 and pr_2 is indistinct (Fig. 33). The second prosomere (pr_2) has enlarged dorsolaterally and forms distinct bilateral spherencephalic buds (*Sp*). The ridge between the pr_2 and pr_3 is especially prominent dorsally and laterally. Its ventrolateral portion constitutes the prospective corpus striatum. The pr_3-sulcus runs laterally and dorsally and fades gradually. The ridge between the pr_4 and the pr_5 is reduced and the two prosomeres merge and form the "pr_{4-5}". The dorsal sulcus within this segment is identical with the sulcus in pr_5 at stage *HH–16*; its ventralmost part extends into the infundibular region, and the two divisions are presumably identical with the sulcus ventralis diencephali and the "sulcus in hypothalamo" of HALLER (1929). A shallow remainder of the pr_5-sulcus still runs into the mammillary recess. The prosomeres pr_{6-8} are distinct. The rostral mesomere has expanded further dorsally to form the optic tectum. Ventrally it is relatively reduced. The m_2 is distinct. The two rostralmost rhombomeres (*Is* and rh_1) are each divided in half on the ventricular surface. The *Is* has a small rostral sulcus (within Is_a) and a larger caudal one (within Is_b). The rh_1 is subdivided only laterally and ventrally (rh_{1a} rostrally and rh_{1b} caudally). This subdivision is much reduced in the succeeding stage (*HH–22*: Fig. 34). The rh_{2-6} are distinct.

The prosencephalic longitudinal sulcus (*S.pr*), present at the preceding stage, begins in the lateral wall of the pr_6 and extends caudalwards, traversing the pr_{7-8} and the m_1 as *S.dl*. Another sulcus (*S.vl*) begins within the "pr_{4-5}" and runs parallel to the aforementioned one, terminating in the rostral mesomere.

The notochord is reduced rostrally in the mesencephalic region, bending ventrally at an acute angle in the isthmic segment. It terminates in the prechordal plate (*pr.ch*).

Stage HH–24. The neural tube has acquired the characteristics of the definite central nervous system (Fig. 35). In the prosencephalon some of the interneuromeric ridges are less conspicuous than at stage *HH–20*, while others are transformed into ventricular swellings. A transverse sulcus is formed in the striate region upon the ridge between the pr_{1-2} and the pr_3. The sulcus within the pr_3 has the same extent and location as in the preceding stage and seems to be identical with the sulcus intraencephalicus anterior of v. KUPFFER (1906). The depth of the dorsal division of the "pr_{4-5}"-sulcus has increased, while the ventral division is shallow. The sulcus is presumably identical with the medial, ventral and hypothalamic sulci of HALLER (1929). The ridge between the "pr_{4-5}" and pr_6 is conspicuous;

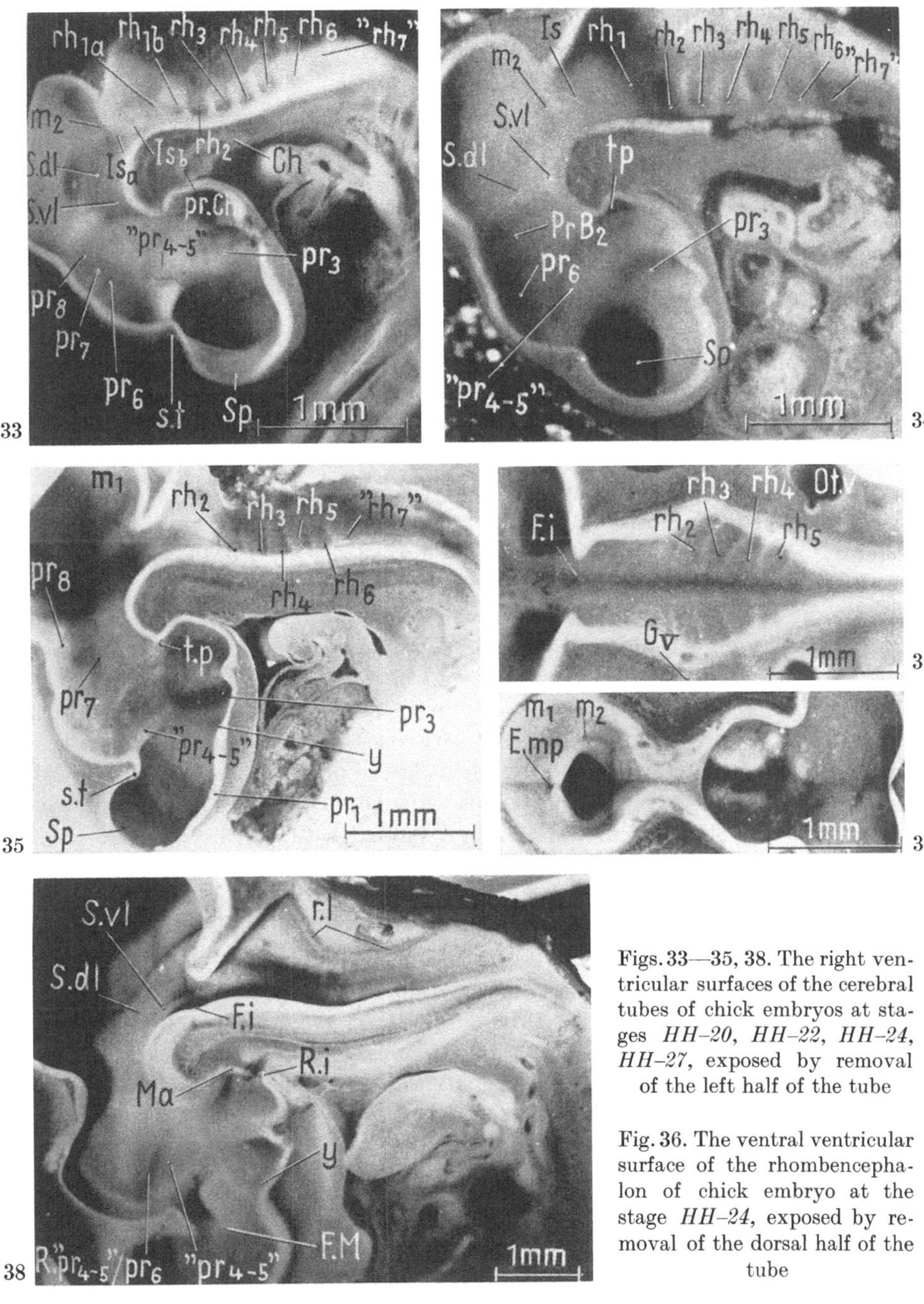

Figs. 33—35, 38. The right ventricular surfaces of the cerebral tubes of chick embryos at stages *HH–20*, *HH–22*, *HH–24*, *HH–27*, exposed by removal of the left half of the tube

Fig. 36. The ventral ventricular surface of the rhombencephalon of chick embryo at the stage *HH–24*, exposed by removal of the dorsal half of the tube

Fig. 37. Frontal view into the isthmic region of chick embryo at stage 24. The roof of mes-, di- and telencephalon is removed

ventrally it is continuous with the posterior tubercle. The pr_6-sulcus is reduced and the central portion of the segment forms a plateau which bulges faintly into the ventricle and represents the rudiment of the dorsal thalamus. The pr_7-sulcus is deepest dorsolaterally and forms the rudiment of the metathalamic recess. The pr_8-sulcus is reduced. In the mesencephalon, the m_1 has expanded to form the tectum opticum. The m_2-sulcus is identical with the posterior intraencephalic sulcus of v. Kupffer (1906) and Palmgren (1921). This sulcus (Fig. 37) begins dorsally in a recess termed the posterior mesencephalic eminence (*E.mp*) by Haller (1929), and runs circularly around the neural tube to terminate ventrally in the rostralmost portion of the isthmic fovea. The ventrolateral and dorsolateral sulci in the pros-mesencephalon are in extent similar to stages *HH–20* and *HH–22*. The two sulci are probably identical with the sulcus limitans and sulcus lateralis mesencephali of Kuhlenbeck (1954). The rhombomeres rh_{2-6} are distinctly developed. Ventromedially in the rostralmost portion of the isthmic segment (Fig. 36) the isthmic fovea of Kingsbury (1922) is recognisable. In the rhombencephalon the square areas described by Bergquist and Källén (1954) are seen (for details, see Vaage, 1969).

Stage HH–27. The neuromeres are further reduced (Fig. 38). The spherencephalon has enlarged dorsally. The sulcus within the pr_3 extends lateralwards into the "pr_{4-5}". The sulcus y along the ridge between the pr_{1-2} and pr_3 runs lateralwards to terminate on the ventral floor of the foramen Monroi. The "pr_{4-5}"-sulcus has grown deeper, separating the rudiments of the ventral and dorsal thalamus. The "pr_{4-5}"/pr_6-ridge is especially distinct and forms a ventricular septum running towards the posterior tubercle. It fades ventromedially in the basal plate region. The ridge becomes still more distinct in the succeeding stages. The infundibular recess has developed in the floor of the pr_4 between the chiasma and the mammillary recess. Several longitudinal sulci are present in the pros-mesencephalon. Dorsally in the pr_6 a sulcus runs caudalwards separating the prospective epithalamus and dorsal thalamus. This sulcus is probably identical with the dorsal diencephalic sulcus of Kuhlenbeck (1954). Another sulcus runs along the "pr_{4-5}"/pr_6-ridge, and ventrolaterally it continues caudalwards into the mesencephalon. The ventrolateral and dorsolateral sulci run from the mesencephalon into the ventrolateral wall of the prosencephalon. The posterior intraencephalic sulcus (within m_2) is conspicuous. It extends, as at preceding stage, into the rostralmost portion of the isthmic fovea. The rh_{2-6} are still discernible, but they are relatively reduced compared with the preceding stage. Several longitudinal sulci extend from the isthmic segment caudalwards into the spinal cord. Within the isthmus, a distinct sulcus occurs. It continues caudalwards and merges with the longitudinal sulcus present at the preceding stage.

Discussion

Figs. 39 and 40 show diagrammatically the development of the segmental pattern as revealed by studies of the superficial and ventricular surfaces of the neural tube in living and fixed chick embryos described above.

A. The Number of Neuromeres

From the three-encephalomere stage, the primary encephalomeres undergo subdivision and transformation as follows:

a) Mesencephalon

The fourth and fifth segments of the cerebral plate constitute, according to HILL (1900) and PALMGREN (1921), the anlage of the mesencephalon. These segments form two mesomeres at stage *HH–9*. No further subdivision of the mesencephalic region is observed. The two mesomeres are most distinctly separated during the early developmental stages, but they are distinguishable throughout the morphogenesis of the brain, as described by v. KUPFFER (1906), MEEK (1907) and PALMGREN (1921). The rostral mesomere enlarges and forms the optic tectum. The caudal one undergoes reduction and forms the boundary segment between the mesencephalon and the rhombencephalon (Figs. 33 and 37).

Some authors have observed only one mesencephalic segment in the later developmental phases (HILL, 1900; NEAL, 1918). These authors described only the rostral mesomere, apparently overlooking the reduced caudal one.

In several papers BERGQUIST, KÄLLÉN and their collaborators likewise described only one mesencephalic segment during the early neurogenesis, *i.e.*, the proneuromere and neuromere periods. In the chick these periods embrace stages *HH–8–9* to *HH–13* (KÄLLÉN, 1956a). During this developmental period the mesencephalon is, according to my interpretation, divided into two segments, and the caudal one begins to fade at stage *HH–13*. Neuromeres IV and V of BERGQUIST and KÄLLÉN (1953a) seem to equal the rostral and caudal mesomeres, respectively, of v. KUPFFER (1906), MEEK (1907), PALMGREN (1921) and the author. Evidently, BERGQUIST and KÄLLÉN have misinterpreted one mesomere in early morphogenesis. Later in the development, after the subdivision of the prosencephalon into *PrA* and *PrB*, but before the subdivision of the PrA_1, "neuromere IV" of BERGQUIST and KÄLLÉN in my opinion is a prosencephalic neuromere. The rostral mesomere is then called "neuromere V" and the caudal mesomere "neuromere VI". A similar interpretation of the neuromeres is found in the papers of BERANECK (1887) and AREY (1965, Fig. 545). However, BERGQUIST and KÄLLÉN (1955) describe a subdivision of the mesencephalon into postneuromeres 6 and 7 at stage *HH–13*.

According to my observations, no subdivision of the mesencephalon takes place during that particular developmental period. By comparing Figs. 1 and 3 of BERGQUIST (1956), neuromere VI (d) in the latter figure evidently is placed further caudally than the neuromere VI (d) in his Fig. 1. The apparent subdivision may, therefore, merely be the consequence of a different interpretation of the bulges in the mesencephalic region at the two developmental stages.

b) Rhombencephalon

The rhombencephalon, in early morphogenesis, is separated from the mesencephalon by the mes-rhombencephalic furrow and from the spinal cord by the rhombo-spinal furrow (Fig. 25). During the early morphogenesis the wall of the rhombencephalon is faintly subdivided into three rhombomeres, *RhA*, *RhB* and *RhC*, which seem to be identical with the prorhombomeres of BERGQUIST and KÄLLÉN (1954, 1955).

At approximately stage *HH–9–10*, the *RhA* and *RhB* subdivide into the RhA_1, rh_3, rh_4 and rh_5, whereby five rhombomeres ($RhA_1 rh_3$–$rh_5 RhC$) are formed.

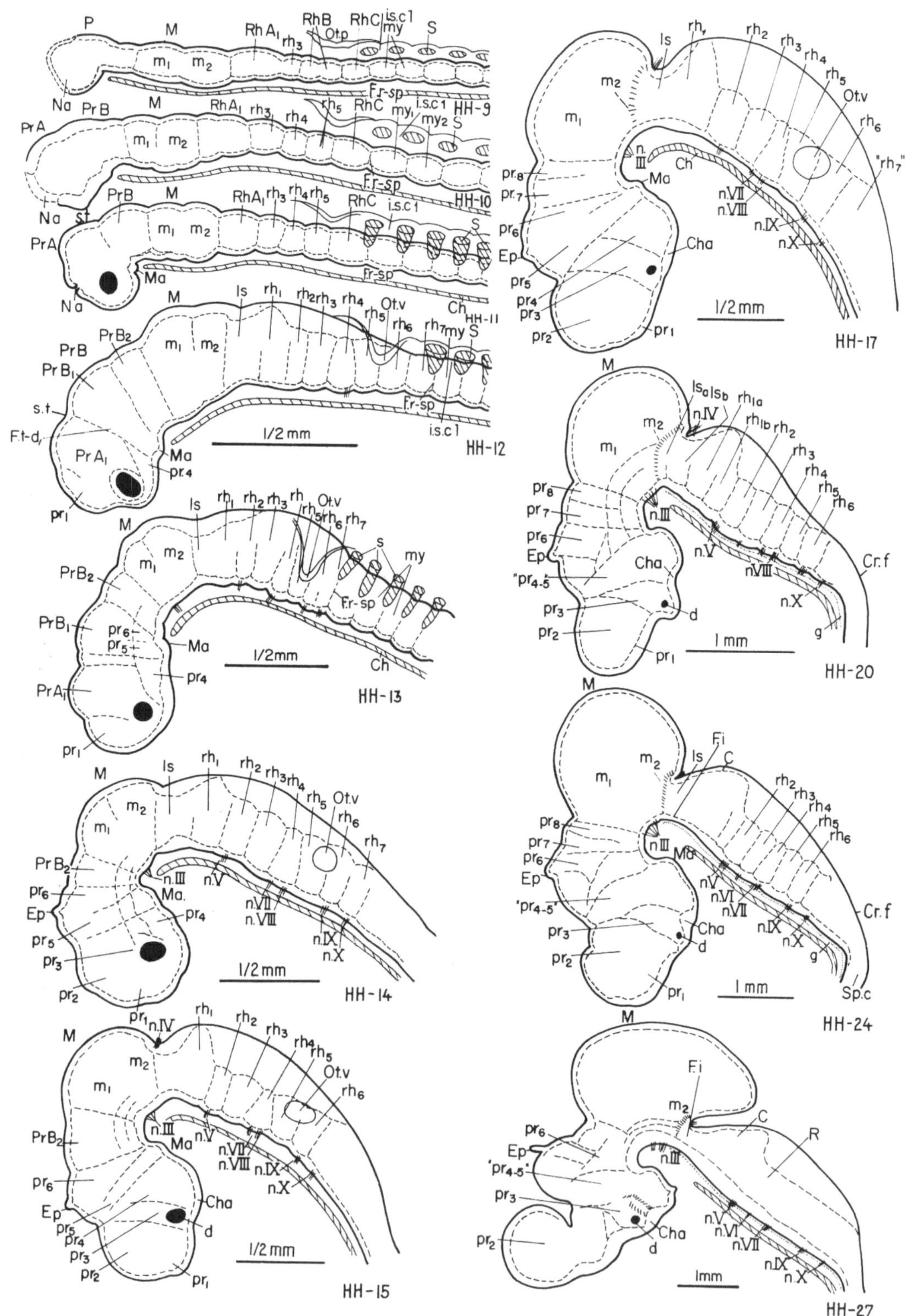

Fig. 39. Diagrammatic drawing of lateral view of chick embryos at stages *HH–9*, *HH–10*, *HH–11*, *HH–12*, *HH–13*, *HH–14*, *HH–15*, *HH–17*, *HH–20*, *HH–24*, *HH–27* showing the development of the neural tube and its relationship to the cranial nerves, the otic vesicle, the notochord and the glycogen-containing raphe

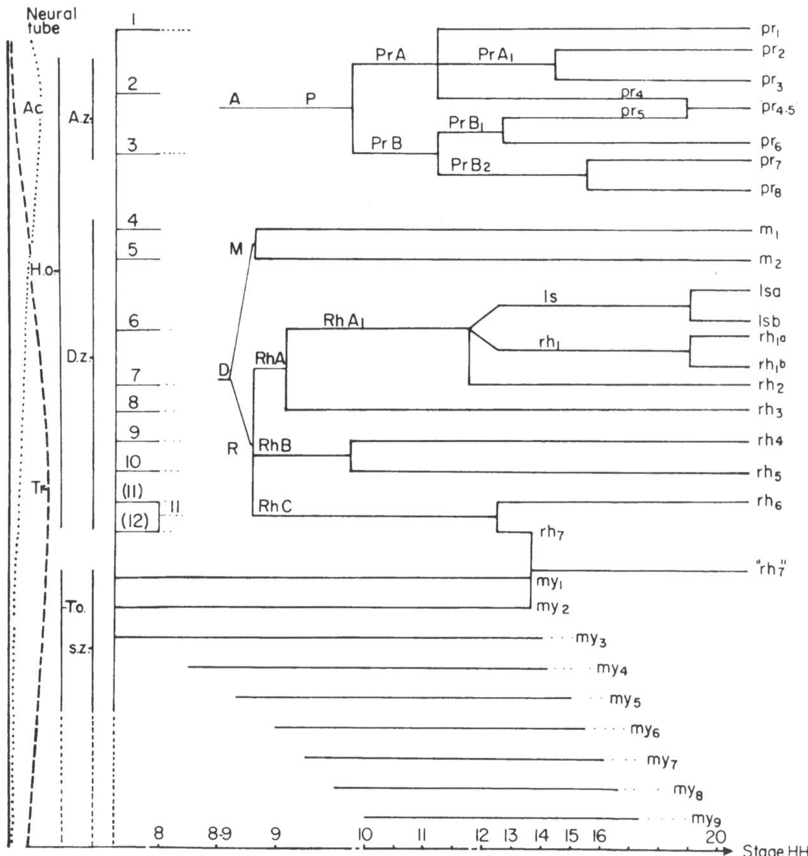

Fig. 40. Diagrams showing the subdivision of the archencephalon (*A*), deuteroencephalon (*D*) and the spinal cord at different stages of development. The relationship between the neural tube and the "primary neuromeres" of HILL (*1–12*), the archencephalic (*A.z*), the deuteroencephalic (*D.z*) and the spinocordal (*s.z*) zone of LEHMANN, the head (*H.o*) and trunk (*T.o*) organizer of SPEMANN, the activation (*Ac.*) and the transforming (*Tr.*) principles of NIEUWKOOP are shown to the left

Identical rhombomeres are repeatedly observed in birds, *e.g.*, by HILL (1900) and NEAL (1918). ZIMMERMANN (1891) and KAMON (1906) also considered the RhA_1 as a metencephalic rhombomere.

BERGQUIST and KÄLLÉN (1953a, 1954, 1955), on the other hand, describe six rhombomeres (VI—XI = d—i) in chick embryos from stage *HH–10* onwards while I observe only five (Figs. 26 and 27). Since they interpret the author's caudal mesomere as the rhombomere VI (d) in the same developmental period (see under Mesencephalon), it appears likely that their neuromeres VII—XI (= e—i) are identical with the author's rhombomeres RhA_1, rh_3–rh_5, *RhC*, respectively. If this assumption is correct, the discrepancy between BERGQUIST and KÄLLÉN's and my conception of the number of rhombomeres at this developmental stage rests on divergent interpretation of the caudal mesomere.

The RhA_1 and RhC subdivide between stages *HH–12* and *HH–13*. In birds, the subdivision of the RhA_1 into the (Is)–rh_1 and the rh_2, is described by Zimmermann (1891), Waters (1892), Neal (1898, 1918), Hill (1900) and Kamon (1906), while the RhC divides into the rh_6 and the rh_7 (Zimmermann, 1891; Waters, 1892 and Neal, 1898, 1918). The seven rhombomeres at this stage are presumably identical with the seven rhombencephalic segments present in the cerebral plate as described by Locy (1895) and Hill (1900). The fifth rhombomere (rh_5) is located medial to the otic vesicle as stressed among others by McClure (1890). It is not connected with any ganglionic cranial nerve (see chapter VI).

Some authors (v. Kupffer, 1906; Bergquist, 1952a, b; Bergquist and Källén, 1953a b, 1954, 1955 and their collaborators) describe only six rhombomeres at comparable developmental stages (see p. 11).

v. Kupffer (1906) published two figures (1906, Fig. 290a, b) comparable with my Figs. 45 and 46, 48—50. v. Kupffer's neuromeres 1—4 are identical with my rh_{1-4}. „Gelegentlich sei bemerkt, daß der Abducens beim Hühnchen dieses Alters (am Übergange vom 3. zum 4. Tage) mit 5 Wurzelbündels aus dem fünften und sechsten Neuromer austritt, der Glossopharyngeus dem fünften und das Wurzelgebiet der Vagus dem sechsten angehört, aber noch darüber hinausreicht" (*l.c.*, p. 264). Neuromeres 5 and 6 in his figures are identical with the rh_5 and rh_6 on my figures. A bulge in the neural wall situated caudal to the sixth neuromere seems to be identical with the "rh_7". The fifth and sixth neuromeres of v. Kupffer accordingly are identical with the rh_{5-6} and rh_7, respectively, in the present study.

The figures published by v. Kupffer do not militate against the interpretations of the present author as shown diagrammatically in my Fig. 39.

Bergquist and Källén (1953a, 1954, 1955) do not describe any increase in the number of rhombomeres after the interval between the first and the second rhombomeric period at stage *HH–10*. Consequently they do not observe the subdivision of either the RhA_1 or the RhC. Their neuromeres VI—XI (= d—i) are, according to my interpretation of the rhombomeres, identical with m_2, RhA_1, rh_3–rh_5, RhC, prior to the subdivision of the RhA_1, with rh_1–rh_5, RhC after the subdivision of the RhA_1 and with rh_1–rh_4, rh_6, rh_7 following the subdivision of the RhC. The otic rhombomere (rh_5) is not recognised at this stage by Bergquist and Källén but is interpreted as the fourth (Källén, 1954, Fig. 1D) or the sixth rhombomere (Bergquist and Källén, 1954, Fig. 2) (see next chapter).

Meek (1907), on the other hand, described two rhombomeres in the same region as the otic rhombomere of the author, *i.e.*, between the rhombomeres which are connected with the facial-acoustic and grossopharyngeal ganglions. The occurrence of only one otic rhombomere is observed by so many investigators and in such a variety of species that it must be considered as the typical pattern.

The rostral rhombencephalic segment, described by Zimmermann (1891) as the sixth segment, may be identical with the isthmus described in chick. Zimmermann's seventh segment may be identical with the rh_1. If this assumption is correct, the isthmic segment described by the author is identical with that described in the chick by Zimmermann (1891).

The caudal rhombomere (rh_7) undergoes reduction at stage *HH–13–14* simultaneously with the reduction of the rostral myelomeres. The rh_7 merges with the rostral myelomeres to form an enlarged rhombomere, the "rh_7" (Fig. 32). A similar "enlargement" of the caudal rhombomere is illustrated by KAMON (1906) and STREETER (1933). The vagus, the spinal accessory and hypoglossal nerves develop according to MINOT (1892) within this enlarged rhombomere.

The division of the segments *Is* and rh_1 occurs relatively late in the neurogenesis (Figs. 33 and 49). This subdivision has previously not been observed, as far as the author knows. The segments Is_a, Is_b, rh_{1a} and rh_{1b} are discernible from *HH–19* to *HH–21*. Their short existence may be one reason why they have been overlooked.

From stage *HH–21–22* onwards, when the segmentation of the *Is* and rh_1 is indistinguishable (Fig. 34), the rhombencephalon rostral to the rh_2 does not appear to be situated within any of the neuromeres. In this developmental period, the rhombencephalon is distinctly subdivided into only six rhombomeres (the rh_{2-6}, "rh_7"). Later on in the development the "rh_7" becomes indistinguishable, and only the five central rhombomeres (rh_{2-6}) are discernible, at least until stage *HH–27* (Fig. 38).

c) Prosencephalon

The prosencephalon consists of one large bulge between stages *HH–8–9* and *HH–10*. At the latter stage it divides into two prosomeres termed *PrA* and *PrB* by v. BAER (1828), MIHALKOVICS (1877), ZIMMERMANN (1891), HILL (1900), KAMON (1906) and PALMGREN (1921).

The interprosomeric boundary between the *PrA* and *PrB* is variously placed. WEBER (1900) and KAMON (1906), using reconstruction models, found that in birds this boundary is situated ventrally between the infundibular anlage and the mammillary recess.

BERGQUIST and KÄLLÉN (1955) named the prosencephalon proneuromere A. It subdivides at stage *HH–10* into two neuromeres called I—III (a) and IV (b). „Die Neuromere a ihrerseits wird von nicht weniger als 4 Postneuromeres gefolgt. Aus der Neuromere a entwickeln sich sowohl Telencephalon als auch Diencephalon (außer des Synencephalon-Bereich). Die Neuromere b bildet den Ursprung zum hinteren Teil des Diencephalon, welcher auch Synencephalon genannt wird, während die Neuromere c lediglich das Mesencephalon bildet" (BERGQUIST, 1964, p. 227). To my knowledge, the segmentation pattern described by BERGQUIST and KÄLLÉN finds little support in the literature. According to my observations, the bulge they describe as neuromere IV (b) prior to *HH–10*, at *HH–10* and at *HH–12* in birds, is identical with the m_1, the *PrB* and the PrB_2, respectively. The bulge described as neuromere I—III (a) at comparable stages is identical with the prosencephalon (*P*), the *PrA* and the $PrA–PrB_1$. If this assumption is correct, the apparent relationship between the neuromere IV and the synencephalon depends upon a misinterpretation of the pros-mesencephalic neuromeres at different ontogenetic stages of development.

The author's observation of the subdivision of the two prosomeres into five (pr_1, PrA_1, pr_4, PrB_1, PrB_2) and shortly later six prosomeres (pr_1, PrA_1, pr_4, pr_5, pr_6, PrB_2) confirms the findings of earlier investigators. The pr_1 is presumably identical with neuromere 1 of WEBER (1900) and neuromerode 0 of BERGQUIST (1952a, b). The pr_{2-4}-anlage is presumably identical with neuromere II of WEBER (1900). The PrB_1 and PrB_2 are presumably identical with the parencephalic and the synencephalic bulges, respectively, observed in birds by Weber (1900), Kamon (1906) and Kuhlenbeck (1954).

The PrA_1 subdivides into pr_2 and pr_3 at *HH–14–15* (Fig. 31). The pr_{1-2} are presumably identical with the bulge called the telencephalon by v. Baer (1828), Mihalkovics (1877), Meek (1907), Rendahl (1924) and Kuhlenbeck (1954). Several authors (Hill, 1900; v. Kupffer, 1906; Meek, 1907, and Grünthal, 1952) describe three prosomeres in the prosencephalon which are identical with the pr_{1-2}, pr_{3-6} and PrB_2. These authors have described the optic stalk within the second prosomere (parencephalon). Hence the tel-diencephalic boundary is placed immediately rostral to the optic stalk and the optic recess. Rendahl (1924) described three diencephalic segments which presumably are identical with the pr_{3-5}, pr_6 and PrB_2.

B. The Pattern of Formation and Reduction of the Neuromeres

The formation and reduction of the neuromeres occur in different ways in the spinal and the cerebral divisions of the neural tube, as shown in Fig. 40.

a) The Spinal Tube

The segmentation starts rostrally in the prospective spinal cord and spreads caudalwards in one wave of bulges (see chapter III) as described by McClure (1890), Zimmermann (1891) and Neal (1918). The myelomeres are thus formed in a rostrocaudal sequence in accordance with the general embryological differentiation pattern shown by Lillie (1919) and Kingsbury (1932). The disappearance of the myelomeres likewise proceeds rostrocaudalwards subsequent to the formation of the spinal nerves, according to McClure (1890) (see chapter III).

b) The Cerebral Tube

From the three vesicle stage (pros-, mes- and rhombencephalon) the formation of segments proceeds in a caudorostral direction. The rhombencephalon starts to subdivide at stage *HH–9* and the prosencephalon at stage *HH–10*.

The subsequent segmentation of the two regions takes place at different stages and proceeds at different rates.

The caudorostral differentiation pattern of the cerebral tube observed in this study essentially confirms the findings of Hill (1900) and Neal (1918). According to McClure (1890) the encephalomeres "degenerate" as soon as the cranial nerves are formed. My observations (see chapter VI) show that the encephalomeres persist for a considerable time even after the formation of cranial nerves.

Bergquist, Källén and co-workers found formation and subsequent disappearance of three different waves of segmentation, *viz.*, proneuromeres, neuromeres and postneuromeres. In other words, encephalomeres form three times during the morphogenesis of the brain, and each time in a rostrocaudal direction. The encephalomeres likewise disappear three times but in a caudorostral direction (Bergquist and Källén, 1954, 1955). According to my observations the encephalomeres begin to disappear after the reduction of the rostralmost myelomeres, and proceed concomitantly in all three brain regions (Fig. 92). In my material the neuromeres could be traced from one bulge into two or more daughter bulges.

Bergquist, Källén and their collaborators state that secondary evaginations develop within some of the rostral bulges during the postneuromery. These evaginations are not transitory formations like the neuromeres. Kuhlenbeck observed a disappearance of the prosencephalic neuromeres and their ventricular grooves which subsequently are replaced by a system of longitudinal sulci (Kuhlenbeck, 1935, 1937, 1938 and 1954).

According to my observations, several remnants of the ventricular sulci persist longer than the neuromery. These sulci can be traced from the neuromeric sulci into the sulci of later ontogenetic stages. Some of the sulci are identified as parts of the sulcus intraencephalicus anterior (pr_3-sulcus) sulcus diencephalicus ventralis ("pr_{4-5}"), medius (rostrally in the pr_6), dorsalis (caudally in the pr_6), recessus metathalamicus (pr_7-sulcus) and the sulcus intraencephalicus posterior (m_2-sulcus). Bergquist (1952a) has observed that the ventricular sulci within the prosencephalon vary at different ontogenetic stages. Thus the sulcus which runs into the optic stalk usually begins within the neuromere II, but in early stages it may begin within the neuromere I. Bergquist interprets this as a variability of the ventricular sulci as well as a migration of the individual sulci, which are of secondary significance and appear concomitantly with various proliferative processes in the neural tube. In view of my observations the sulci within the neuromere I at early stages and within the neuromere II at later stages are identical with the sulci within PrA_1 and pr_3, respectively. The sulci have not migrated on the ventricular surface, but the sulcus within the pr_3 is formed by subdivision of the PrA_1.

C. The Relationship between Neuromeric and Permanent Sulci

McClure (1890) found encephalomeres in chick between 36 and 96 hours of incubation, *i.e.*, between stages *HH–11* and *HH–24*. Kuhlenbeck (1954) looked upon the neuromeres as transitory bulges. The diencephalic neuromeres disappear shortly after their formation. According to my observations some of the neuromeric sulci in chick can be traced stage by stage without the preceding transformation and disappearence into permanent sulci, *e.g.*, sulcus intraencephalicus anterior and posterior, s. thalamicus ventralis, medius and recessus metathalamicus of Haller.

Bergquist and Källén (1953a, b, 1954, 1955) describe the occurrence of three or even four successive waves of ventricular sulci varying both in location and morphology. Hence the sulci cannot be used as landmarks during the early ontogenesis (Bergquist, 1964). In my material some of the sulci may be traced continuously from their early beginnings and to their final positions in the brain wall.

Conclusion

New encephalomeres are formed by successive subdivisions of the preceding neuromeres. In this way eight prosomeres, two mesomeres and eight rhombomeres are created.

The cerebral tube differentiates initially in a caudorostral direction, the spinal tube in a rostrocaudal direction. Later on the prosencephalon, the rhombencephalon and the spinal cord subdivide independently of each other.

Some of the neuromeric sulci develop into intracerebral sulci.

VI. Observations on Serial Sections

As shown in the preceding chapters the neural tube during its ontogenetic development forms bulges and furrows on the superficial surface and sulci and ridges on the ventricular surface. Serial sections are necessary in order to ascertain that the bulges display all the characteristics of the neuromeres as defined by ORR (1887). Furthermore serial sections enable detailed study of the relationship between the various structures which are observed.

Results

Stage HH–8. The "primary neuromeres" of HILL (1900) are discernible (Fig. 41). The segments are especially distinct at intersomitic levels of the spinal cord anlage and are separated by ridges and corresponding superficial furrows. Of the four somites the rostral one is incompletely separated from the head mesoderm. Therefore it is called a rudimentary somite. The others are separated by distinct intersomitic clefts.

Stage HH–9. The entire neural tube is composed of bulges (Figs. 42 and 70) like those observed in living and dissected specimens. In the neural groove region the lateral halves are segmented as in the preceding stage. Superficial bulges correspond to ventricular sulci, and the furrows correspond to ventricular ridges. An optic evagination is seen rostrally in the archencephalon (Fig. 42) confirming the observations on living specimens (p. 19). The rh_3-sulcus is presumably homologous to the preotic sulcus described by ADELMANN (1925) and BUTCHER (1929). The notochord (Fig. 70), which is compact, corresponds to the caudal half of the deuteroencephalon. Rostrally the notochord forms a relatively loose string of cells, terminating at the arch-deuteroencephalic boundary. In the cerebral tube the neural folds approach each other middorsally. A distinct cell mass, evidently the neural crest, occupies the angle between the ectoderm and the neural wall in the caudal prosencephalon (Fig. 54), in the prospective mesencephalon (Fig. 55) and in the rostral part of the rhombencephalon (Figs. 56 and 57). This fully corroborates previous observations (p. 21). The neural crest is lacking in the caudal half of the rhombencephalon (Fig. 58).

Stage HH–10. The anterior neuropore is reduced to a sagittal split, while the caudal one is open. The optic evagination has grown further laterally (Fig. 43). Bulges and furrows and corresponding ventricular sulci and ridges like those observed in living and dissected specimens are present. Four rhombomeres RhA_1, rh_3, RhB and RhC, are discernible. The RhB is narrow, laterally faintly subdivided into a rostral (rh_4) and a caudal (rh_5) rhombomere. No subdivision of the rostral myelomeres can be detected. The otic placode (*Ot.p*) is demarcated dorsolateral to the RhB as in corresponding living specimens (p. 21). The neural crest extends antero-posteriorly from the prosencephalon, through the mesencephalon (Fig. 59) and the rhombencephalon (Fig. 60) into the rostralmost portion of the spinal cord (Fig. 62). Dorsal and lateral to the rh_3 the neural crest is thinning out, thus forming a rostral (pros-rhombencephalic) and a caudal (rhombencephalic) division. The pros-rhombencephalic neural crest has a

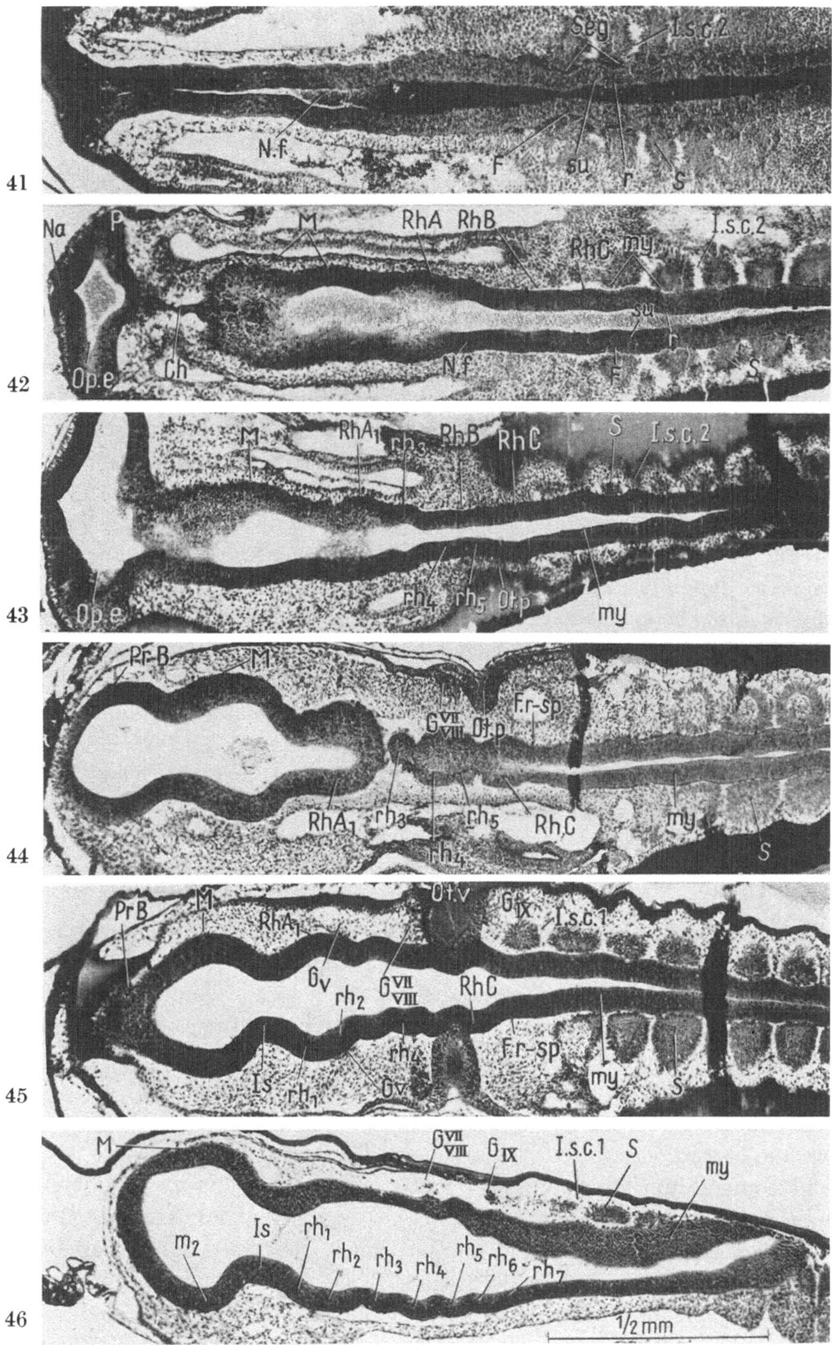

Figs. 41—46. Longitudinal sections through the neural tubes of chick embryos at stages *HH-8*, *HH-9*, *HH-10*, *HH-11*, *H-12*, *HH-13* showing the neuromeres at different stages of development

lateral distribution similar to the crest observed *in vivo* at stages *HH–9–10* and *HH–11* (p. 21). The rhombencephalic neural crest is especially crowded with cells just rostral and caudal to the otic placode. The two divisions are connected by a lose string of cells mid-dorsally (Fig. 61).

Stage HH–11. The neuromeres are essentially identical with those present at stage *HH–10* (Fig. 44). The prospective transverse septum is discernible dorsally between the *PrA* and *PrB*. The floor of the mesencephalon (Fig. 71) is compressed in a rostrocaudal direction. The resulting transversal sulci are without morphological significance, according to Orr (1887). The superficial two-thirds of the wall of the neural tube are densely packed with epithelial cells, while in the ventricular one-third the cells are more loosely arranged (Figs. 63—69). The notochord (Fig. 71) is more solid than at stage *HH–9*. The pros-rhombencephalic neural crest is attenuated rostrally, and laterally the cells mingle with the head mesoderm. At the level of the *RhA* (Fig. 63) the crest is condensed and forms the rudiment of the semilunar ganglion. No neural crest is present corresponding to the rh_3-rhombomere (Fig. 64). The preotic (Fig. 65) and the post-otic (Fig. 67) portions of the rhombencephalic neural crest, interconnected by a mid-dorsal cell string (Fig. 66), are very compact and form the prospective facial-acoustic and petro-nodose ganglions, respectively. The postotic neural crest continues uninterrupted into the spinal neural crest (Fig. 68). The neural crest has not yet formed caudally in the spinal cord (Fig. 69). The opaque structures observed in living embryos have the same relationship to the neural tube as the pros-rhombencephalic, the preotic and postotic portions of the neural crest and are presumably identical structures.

Stage HH–12. The subdivision of the rhombomere RhA_1 observed in living specimens is confirmed (Fig. 45). The segments have no corresponding superficial bulges, and the Is-segment has no ventricular sulcus. The rhombomeres rh_3–RhC are unaltered. The myelomeres occur throughout the somitic portion of the spinal cord. The otic placode forms a pit (*Ot.v*) lateral to the rh_5 and the *RhC*, and occupies the superficial furrow between them (Fig. 45). The notochord (Fig. 72) has grown still more solid than at the preceding stage, but its caudorostral extent is unaltered. The mes-rhombencephalic subdivision of the deuteroencephalon is epichordal, while the prosencephalon (archencephalon) is prechordal. The neural crest is essentially the same as in the preceding stage.

Stage HH–13. The three prosomeres *PrA*, PrB_1 and PrB_2 observed on partly sectioned specimens correspond to ventricular sulci. They are separated by furrows and corresponding ventricular ridges. The histological structure is similar in the prosomeres and the rhombomeres (Fig. 46) and displays the characteristics of the neuromeres as defined by Orr (1887). The m_2 (Fig. 46) is restricted to the caudal half of the mesencephalon. The previously described (p. 38) subdivision of the *RhC* into the rh_6 and rh_7 is distinct. Experiments prove that no rostral somites disappear in the developmental period between *HH–10* and *HH–14* (Vaage and Høivik, 1969). The two rhombomeres (rh_6 and rh_7) interposed between rh_5 and the rostralmost myelomere must therefore differentiate within the *RhC*. In the spinal cord, the rostral myelomeres are somewhat reduced, while the myelomeres located further caudally are distinct. The mid-dorsal neural crest of

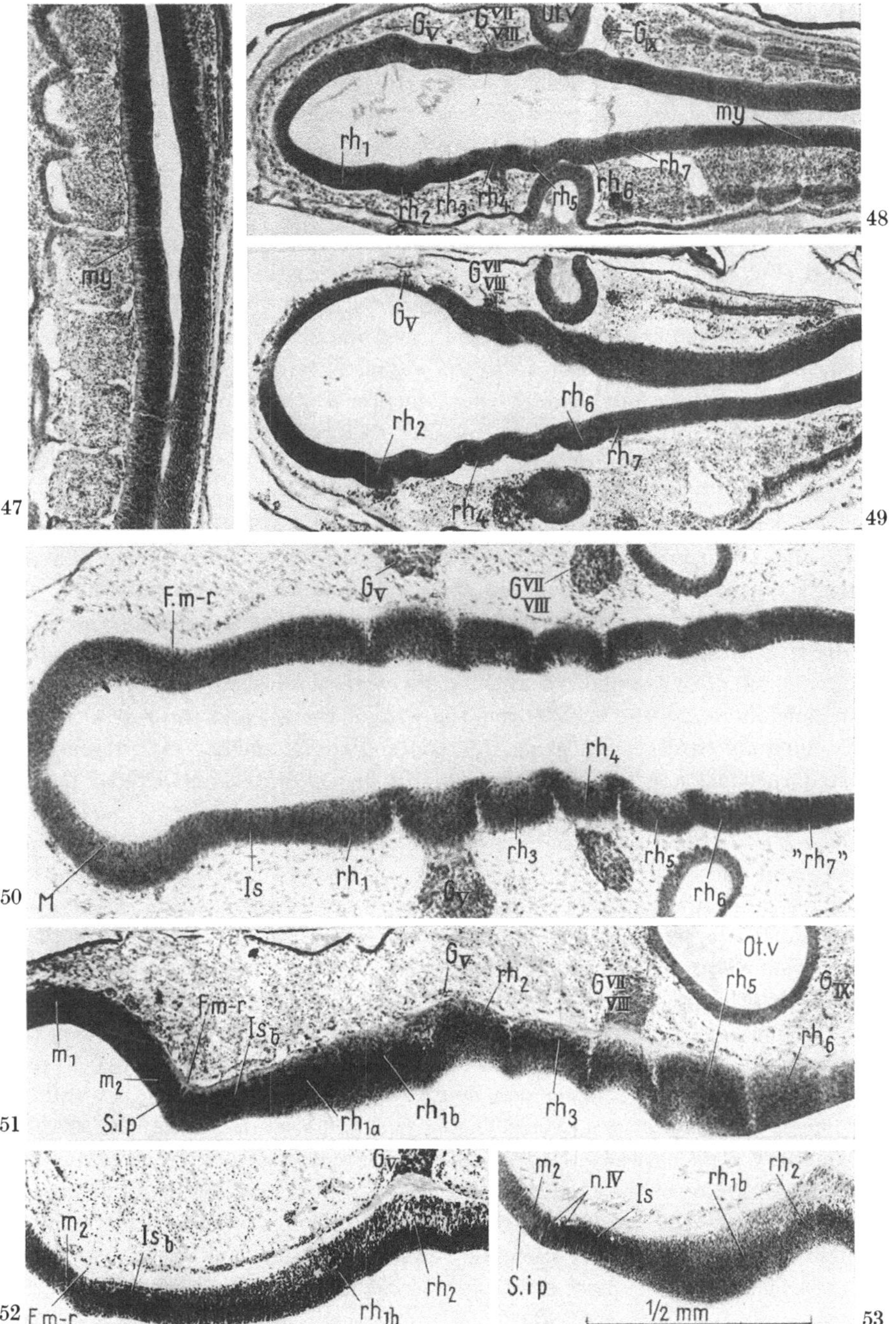

Figs. 47—53. Longitudinal sections through the spinal cords (Fig. 47) and rhombencephalons (Figs. 48—53) of chick embryos at stages *HH-14* (Fig. 48), *HH-15* (Figs. 47, 49). *HH-17*, *HH-20*, *HH-22*, *HH-24* showing the neuromeres at different stages of development

the rh_5 has disappeared, leaving the pre- and postotic parts completely separated. The postotic neural crest is continuous with the neural crest of the spinal cord.

Stage HH–14. The PrA_1 observed on partly sectioned specimens has enlarged dorsally. Ventrally the optic evagination is constricted to form an optic stalk. Remnants of the furrow $F.t\text{–}d_1$ and its corresponding ridge, previously described (p. 29), are discernible between the PrA and PrB_1. The narrow isthmic segment has its greatest length ventrally, while the rh_1 has its greatest extent dorsally and constitutes the rudiment of the cerebellum. The rostral myelomeres have vanished (Fig. 48) while the more caudal ones are distinct.

Stage HH–15. The serial sections fully confirm the findings in living, dissected and partly sectioned embryos. As at the preceding stages, the dorsal half of the prosencephalon is subdivided into three prosomeres, termed PrA, PrB_1 and PrB_2 (Fig. 39). Further ventrally in the prosencephalon several dorso-ventrally running sulci occur, subdividing the prosomeres into lesser segments. Histologically the four prosomeres pr_{1-4}, located within the PrA, are visible as wide sulci separated by ridges (Fig. 39). A superficial furrow ($F.t\text{–}d_2$), presumably identical with the tel-diencephalic sulcus of MEEK (1907), HOCHSTETTER (1919) and KUHLENBECK (1954), and a corresponding ventricular ridge separate the pr_2 from the pr_3. Ventrally the boundary between the PrA and PrB has become indistinct. The pr_5 and pr_6 observed in partly sectioned specimens are well-defined. The pr_5 is dorsally larger than the other prosomeres. The rh_7 is less distinct than in preceding stages (Fig. 49). The rhombo-spinal ridge, its corresponding furrow and the rostral myelomeres have vanished. However, the wall of the rh_7 and the rostral myelomeres bulge outwards constituting the enlarged rhombomere "rh_7" observed in dissected specimens (p. 32). The myelomeres in the central and caudal portion of the spinal cord are conspicuous (Fig. 47).

Stage HH–17. The prosomeres are more well-defined. The pr_2 bulges dorsolaterally and the ridge between it and the pr_3 is more prominent than earlier. The pr_3, forming the half of a segment on each side of the tube, begins dorsally at the transverse septum (Fig. 39) and runs ventralwards to coalesce with its opposite half ventrally. The optic stalk originates from its ventrolateral portion (Fig. 39), and the optic recess is located ventrally. The m_2-sulcus is distinct around the entire neural tube. The Is and rh_1 are separated by a ventricular ridge (Fig. 50). Both segments have a shallow ventricular sulcus. The "rh_7" has enlarged laterally and occupies the same region of the neural tube as the rh_7 and the rostralmost myelomeres at earlier stages. Neuroblasts are aggregated in the mantle layer. At the boundary zone between the rhombomeres few neuroblasts are present. This relatively cellfree space is a limiting zone between neighbouring neuromeres.

Stage HH–20. The subdivision of the prosencephalon is less distinct than at the preceding stages. Several sulci, as well as intersegmental ventricular ridges and superficial furrows, present in the ventral half of the procencephalon, are reduced, while the PrA, PrB_1 and PrB_2 are discernible dorsally. The pr_2 bulges out dorsolaterally (Fig. 39) forming the prospective spherencephalon observed in partly sectioned specimens. The ventricular ridge corresponding to the $t\text{–}d_2$-furrow is prominent and divides the PrA into a rostral (pr_{1-2}) and a caudal (pr_{3-4}) portion. The border between the pr_4 and pr_5 has vanished, and the two segments in question coalesce to form an enlarged segment, termed "pr_{4-5}", with one ventricular sulcus (Fig. 39). The pr_3 starts medially in the optic recess

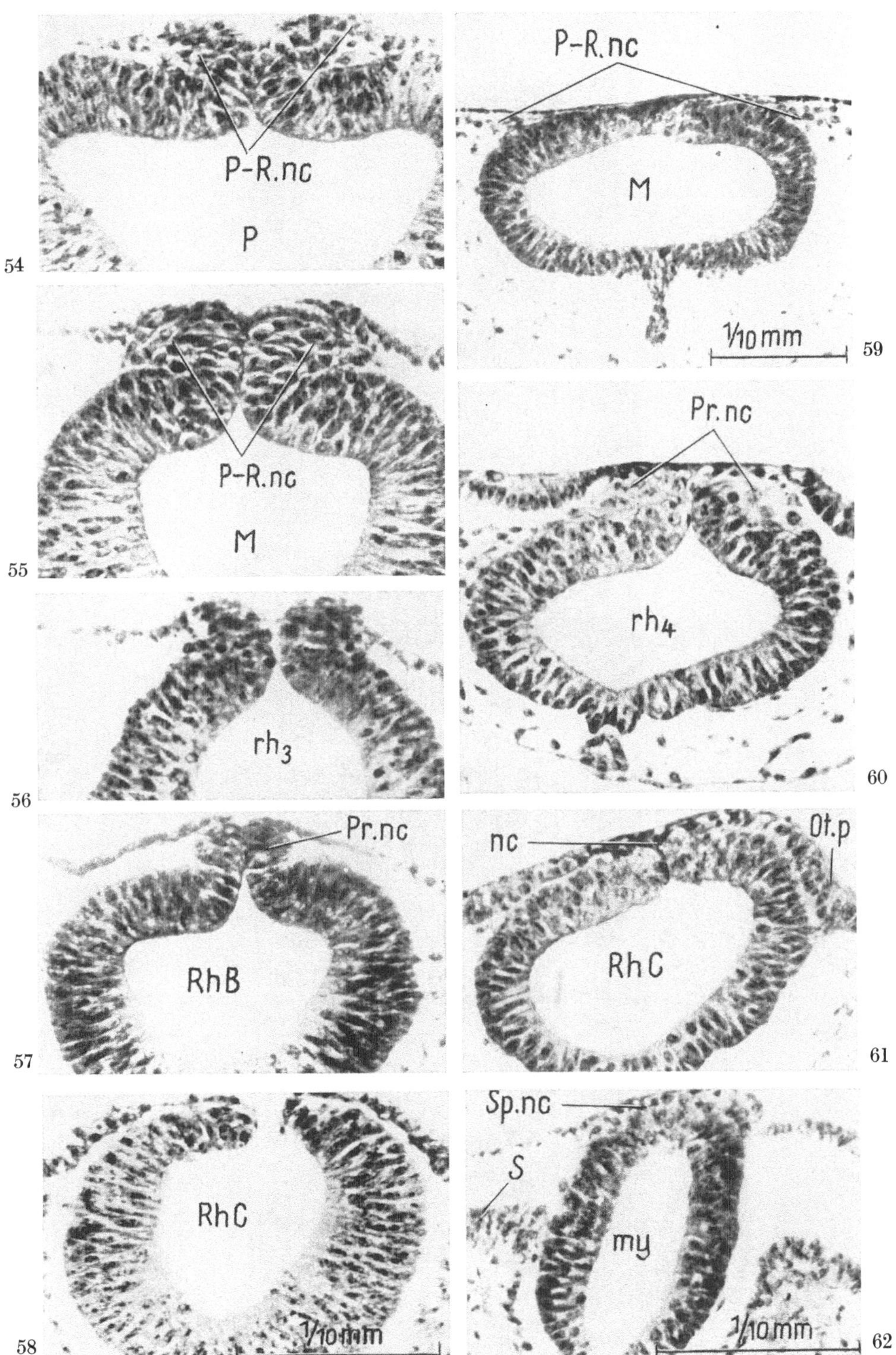

Figs. 54—58. Transverse sections through the prosencephalon, mesencephalon, rh_3, *RhB* and *RhC* of chick embryo at stage *HH-8-9*

Figs. 59—62. Transverse sections through the mesencephalon, rh_4, *RhC* and the spinal tube of a chick embryo at stage *HH-10*

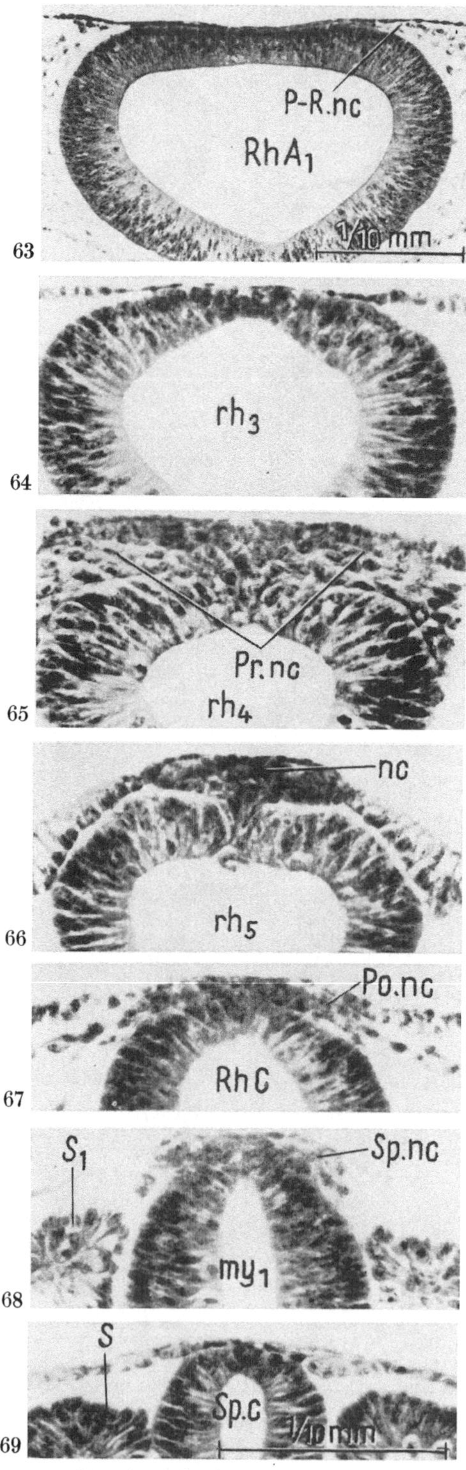

Figs. 63—69. Transverse section through the RhA_1, rh_3, rh_4, rh_5, RhC, my_1 and further caudally in the spinal tube of the chick embryo at stage *HH–11*

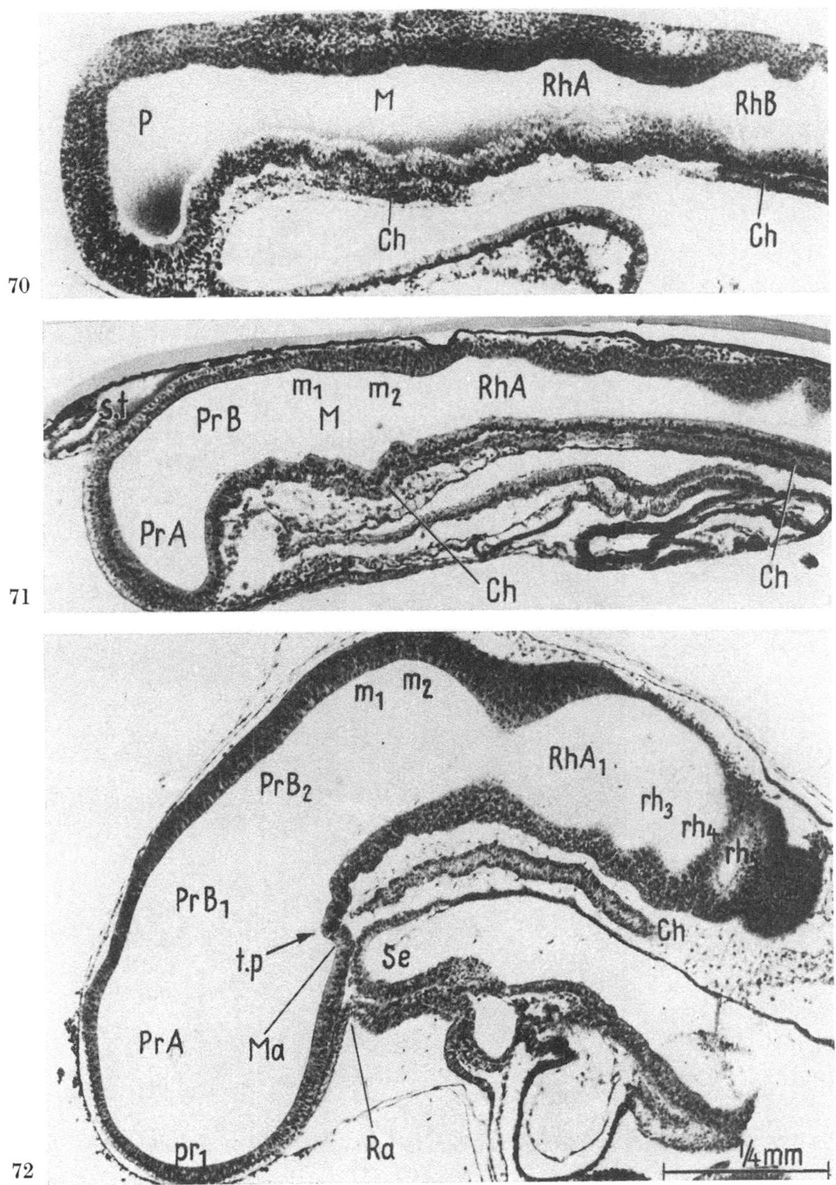

Figs. 70—72. Sagittal section through the cerebral tubes of chick embryos at stages *HH-9*, *HH-11*, *HH-12*

and runs lateralwards to terminate gradually in the lateral wall of the "pr_{4-5}". The pr_6 possesses dorsally the rudiment of the epiphysis. The pr_7-sulcus is distinct, while the pr_8-sulcus is somewhat reduced. The pr_7 and pr_8 are identical with the areas precommissuralis and commissuralis, respectively, of RENDAHL (1924). The rostral mesomere has expanded dorsally to form the prospective optic tectum; ventrally it is reduced. The caudal mesomere is reduced to a dorsoventral sulcus on the ventricular surface; its mantle is cellfree (Fig. 51). The serial sections

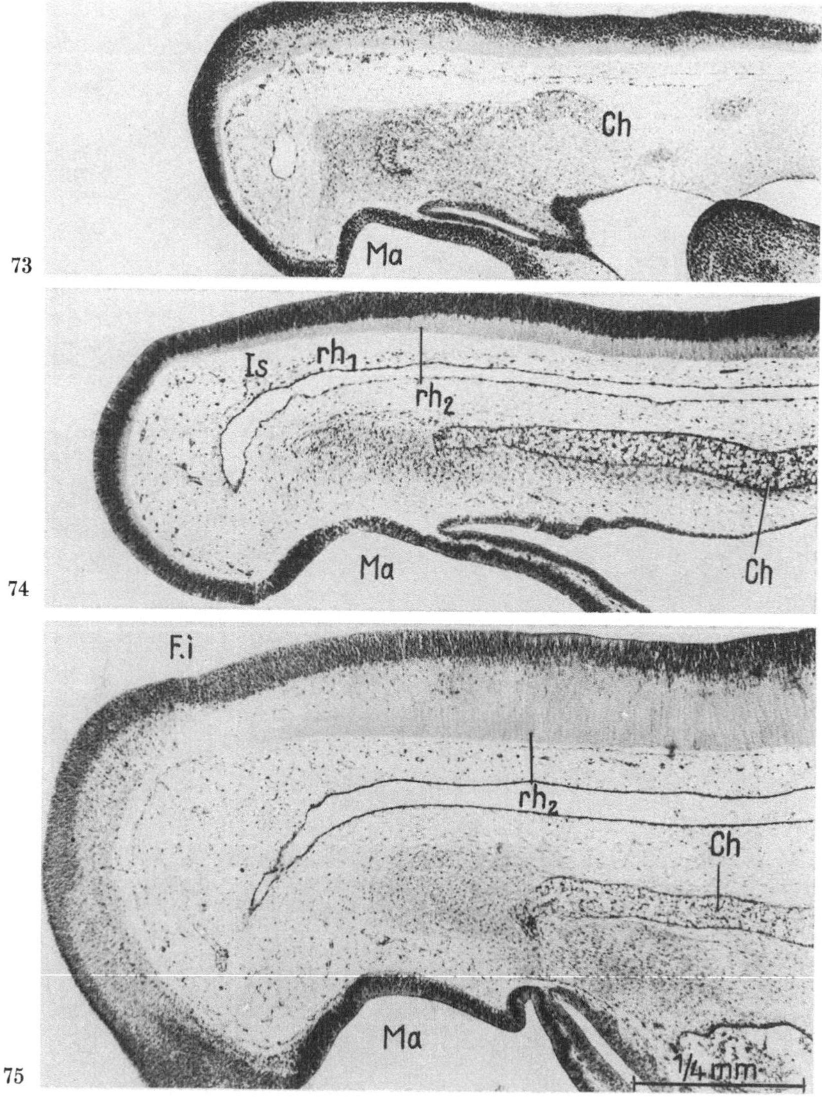

Figs. 73—75. Sagittal sections through the cerebral tubes of chick embryos at stages *HH-20*, *HH-24*, *HH-27*

corroborate the subdivision of the *Is* and the rh_1 observed on partly sectioned specimens (Fig. 51). The Is_a, Is_b, rh_{1a} and rh_{1b} are separated by ventricular ridges running dorsoventrally. No corresponding superficial furrows are seen. In the succeeding stages the subdivision of the Is and the rh_1 is less well-defined (Fig. 52). The rhombomeres rh_{2-6} as well as the caudal myelomeres occur. The oculomotor, trochlear, trigeminal, facial-acoustic, glosso-pharyngeal and vagus nerves emerge from the m_1, *Is*, rh_2, rh_6, and the "rh_7", respectively. The characteristics features of the prosomeres and the mesomeres are less conspicuous. The neuroblasts

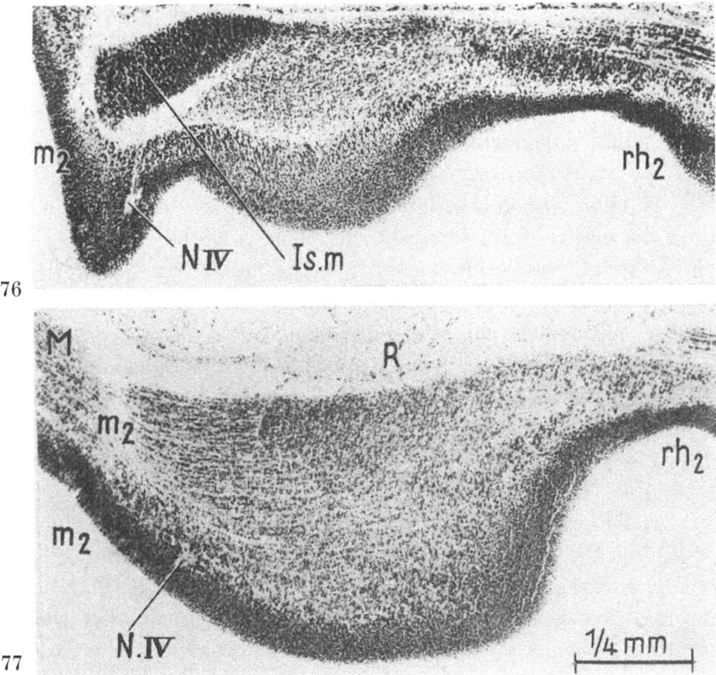

Figs. 76 and 77. Longitudinal sections through the rostralmost portion of the rhombencephalon of chick embryo at stage *HH–27*

aggregate in the mantle in both the cerebral and the spinal tube. The notochord extends rostralwards to terminate in the prechordal plate in the mammillary region (Fig. 73). Two longitudinal sulci occur within the lateral wall of the pros- and mesencephalon (Fig. 39).

Stage HH–24. The reduction of the neuromeres between stages *HH–20* and *HH–24* observed on partly sectioned specimens is ascertained by histological examination. However, the transformation of the neuromeres into ventricular swellings and sulci can be traced stage by stage. The ridge between the "pr_{4-5}" and the pr_6 has become more prominent and is similar in extent to the ridge observed previously (p. 42). The dorsal half of the ridge is cellfree and serves as a landmark between the rudiments of the ventral (pr_{4-5}) and the dorsal (pr_6) thalamus. The sulci within the pr_3, the "pr_{4-5}" and the pr_7 are identified as the sulcus intraencephalicus anterior of v. Kupffer (1906), sulcus thalamicus ventralis and medius and sulcus metathalamicus of Haller (1929). However, the sulci are separated from the surroundings by indistinct swellings. The caudal portion of the pr_6 bulges into the ventricle and forms the prospective dorsal thalamus. Its rostral sulcus has attained a more rostral position just caudal to the "pr_{4-5}"/pr_6 ridge, while its caudal sulcus extends into the pr_7-sulcus (Fig. 39). The m_2-sulcus is distinct (Fig. 53) and identical with the sulcus intraencephalicus posterior of v. Kupffer (1906). The boundaries between the Is_a, Is_b and rh_{1a} have vanished. The sulci within the rh_{1b} and the rh_{2-6} are discernible, while the superficial furrows between them are reduced. The oculomotor and trochlear nuclei are located in the floor of the rostral mesomere and the isthmus (*Is*), respectively. The cellfree mantle of the m_2 is interposed between the two pairs of sulci. The notochord, terminates in the mammillary region below the floor of the rh_1 (Fig. 74). The mantle layer located medial and lateral to the ventrolateral mesencephalic sulcus corresponds to the ventral and dorsal mesencephalic area, respectively, of Bergquist and Källén.

Stage HH–27. The transformation of the neural tube has continued, and the neuroblasts in the mantle form several layers. The ridge between the "pr_{4-5}" and pr_6 is still more distinct vanishing ventrally in the region of the posterior tubercle. The sulci within the pr_3, "pr_{4-5}", pr_7 and m_2 are discernible (Figs. 76, 77). The lateral halves of the caudal mesomere are displaced rostralwards together with the trochlear nucleus and the isthmic segment. The displacement is even more pronounced in the subsequent developmental period. The caudal mesomere with its cellfree mantle is interposed between the rostral mesomere and the isthmus (Figs. 76, 77). The ventricular sulci within the rhombomeres are further reduced, but still discernible. The notochord terminates in the mammillary region below the rh_2 (Fig. 75). In sagittal section (Fig. 75) the ventral wall of the isthmus is depressed, forming the isthmic fovea observed on partly sectioned specimens. The floor plate is distinguished from this fovea caudalwards. The longitudinal sulci within the pros-mesencephalon and the rhombencephalon observed on partly sectioned specimens are discernible. Within the rhombencephalon several longitudinal sulci are discernible.

Discussion

A. The Neuromeres

a) The Histology of the Neuromeres

My results lend support to the findings of McClure (1890) which showed that the histological structure of the encephalomeres at early developmental stages display the characteristics of the neuromeres as described by Orr (1887). Later only some of the encephalomeres undergo reduction while others persist, separated by cellfree zones (Figs. 50, 51). Neal (1898, 1918), studying various vertebrates, pointed out a striking structural and morphological difference between neuromeres in different regions of the neural tube. Hence he rejected the hypothesis that the rhombomeres are serially homologous with the anteriorly and posteriorly located segments. Streeter (1908, 1911, 1933) and Bartelmez (1923) among others, observed neuromeres only within the rhombencephalon. Källén (1954), on the other hand, describes the segments observed throughout the neural tube as serially homologous neuromeres. In view of my observations at early stages of development, the encephalomeres and the myelomeres have a similar histological structure. However, the myelomeres appear and disappear simultaneously with the appearance and disappearance of the adjacent somites. The encephalomeres are profoundly transformed during the early ontogenesis simultaneously with the migration of neuroblasts into the mantle layer.

b) The Segmental Significance of the Neuromeres

Some authors (McClure, 1890; Zimmermann, 1891; Hill, 1900; Kamon, 1906; Bergquist and Källén, 1954) maintain that the neuromeres are segments of the neural tube. Others deny the existence of neuromeres except in the rhombencephalon or explain the neuromeres as caused by mechanical factors (MIHALKOVICS, 1877; NEAL, 1898). These authors stressed the variability of the neuromeres and the fact that many investigators were unable to find archencephalic neuromeres. The rhombomeres do exist but lack segmental significance. My observations show that similar segments occur at early stages in both the archencephalon, the deuteroencephalon and the spinal tube. The segments are structurely identical with the neuromeres as defined by ORR (1887). Hence they are termed neuromeres. In the cerebral tube the neuromeres occur prior to the development of nerves, and the myelomeres occur concomitantly with the somites.

c) The Causes of the Neuromeres

Previous investigators have suggested several causes for the neuromery. BARTELMEZ (1923) pointed out that the formation of neuromeres is preceded by focal proliferation of the neural epithelium. COGHILL (1924) demonstrated the existence of proliferation centres in the neural tube of amphibians. It remained, however, for KÄLLÉN (1952) to prove that the proliferation centres in the neural epithelium coincide in position with the encephalomeres. Each neuromeric sulcus represents a proliferation centre. As the proliferation increases, mitoses accumulate along the neuromeric sulcus, and the bulge widens. If, on the other hand, the proliferation decreases, the bulge becomes narrower (KÄLLÉN, 1953).

According to my observations, the prosomeres and rhombomeres are formed by subsequent subdivision of preceding neuromeres. If BARTELMEZ (1923) and KÄLLÉN (1952, 1953) are correct, each subdivision and formation of new encephalomeres is preceded by formation of new proliferation centres at the particular level of the neural tube. This concept implies the existence of several maxima of proliferation in encephalomeres which subdivide several times. In Fig. 78 the subdivisions of *RhA* and the corresponding proliferation activity are outlined. The *RhA* subdivides into rh_3 (3) and RhA_1 (A) between *HH-9* and *HH-10*. The latter subdivides into rh_2 (2) and $Is\text{–}rh_1$ (0—1) at *HH-12*. At stage *HH*-15 the $Is\text{–}rh_1$ is separated into *Is* (0) and rh_1 (1), both subsequently subdividing twice at stage *HH-19–20*. Four proliferation maxima, occuring just prior to the formation of the *RhA* [*HH-9*, (1), Fig. 78b], rh_2 (2) *Is* and rh_1 (3), and Is_a, Is_b (4) presumably exist if this hypothesis is correct.

KÄLLÉN (1956a) mapped the proliferation activity in chick embryos by means of the relative mitotic rate. The investigation was performed in the "rostralmost end of the rhombencephalon. This part was easily identifiable in all stages" (l.c., p. 177). This area is presumably identical with the *RhA*-region. KÄLLÉN revealed three (a—c, Fig. 78c) proliferation maxima and possibly a fourth one (×, Fig. 78c) coinciding in age with the four maxima (1)—(4) indicated in Fig. 78b. In my opinion it appears reasonable to correlate the variation of the proliferation activity observed by KÄLLÉN with the subdivision of the rhombomere and its derivatives. BERGQUIST (1957a, b) investigated the relative mitotic rate of the neural epithelium in the area caudalis thalami of chick, which is presumably identical with the lateral wall of pr_{7-8}. A proliferation maximum exists between stages *HH-18* and *HH-20*. Compared with my findings, the maximum occurs simultaneously with the subdivision of the RhB_2. It still must be proved that

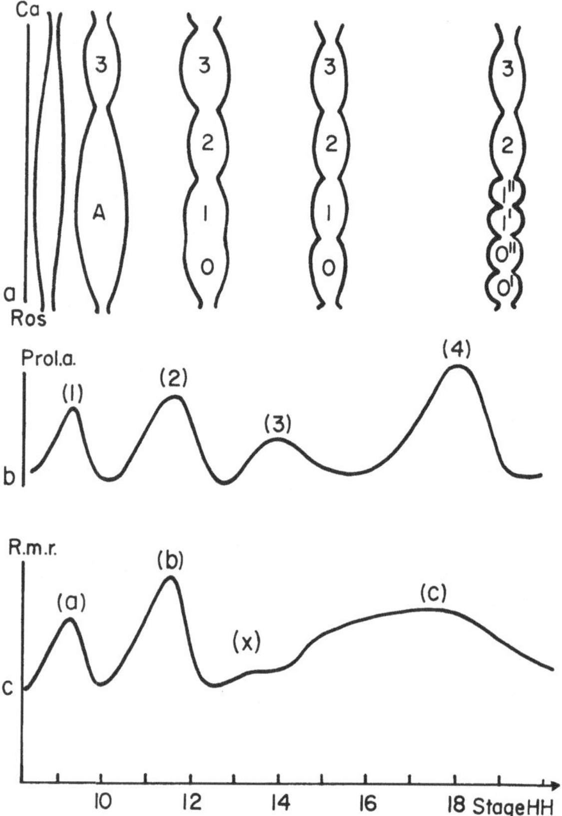

Fig. 78a—c. Diagrams showing the relationship between the formation of neuromeres and proliferation activity. a The subdivision of the *RhA*. b The hypothetical proliferation variations in the neural epithelium within *RhA*. One proliferation peak (*1–4*) is expected to occur prior to each subdivision of an encephalomere. c Diagrams showing variations with age in the relative mitotic rate in the rostral end of the rhombencephalon in chick embryos (Modified from KÄLLÈN, 1956a)

there is a causative relationship between proliferation maxima and neuromere formation.

A reinvestigation of the proliferative activity in neuromeres with different patterns of division is indispensable for solving this problem. Bulges which do not subdivide (4, Fig. 79a) or subdivide once (3), twice (2), three (1) or four (0) times should be investigated. If the variation of the proliferation activity in the neural epithelium occurs only prior to neuromere formation, the abovementioned types of neuromeres must have none (Fig. 79b), one (Fig. 79c), two (Fig. 79d), three (Fig. 79e) or four (Fig. 79f) proliferation maxima, respectively.

B. The Relationship between Several Secondary structures and the Neuromeres

a) The Notochord

The notochord develops during the very early period of ontogenesis, as previously shown by v. BAER (1828), FASTER and BALFOUR (1876), MIHALKOVICS

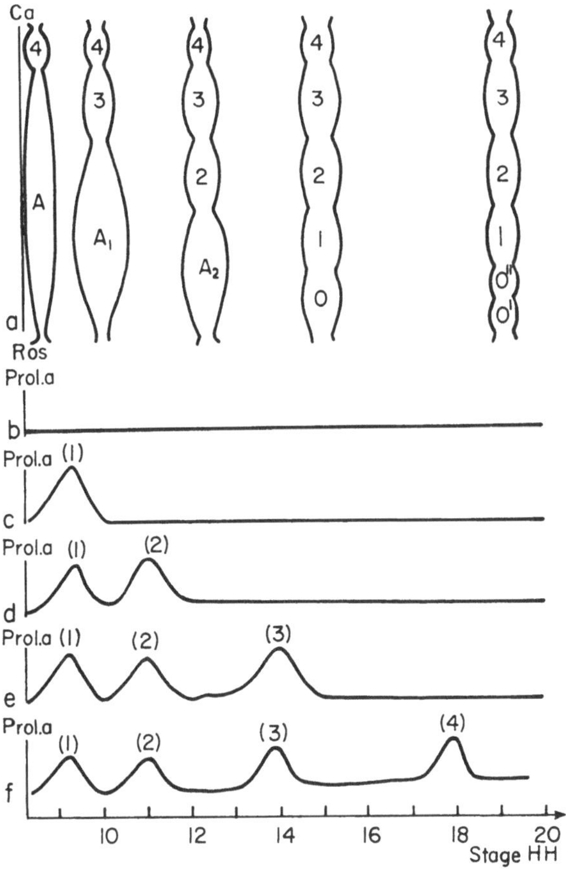

Fig. 79a—f. Diagrams showing the hypothetical relationship between the formation of neuromeres (a) and variation in the proliferation activity in the neural epithelium at different levels of the neural tube (b—f)

(1877), Hill (1900), Lillie (1919), Kingsbury (1924), Hillemann (1943), Lehmann (1945), Spratt (1947, 1952) and Patten (1958). In early developmental stages the notochord is especially distinct below the rhombencephalon and the rostral part of the spinal cord. Rostrally the notochord is continuous with a slender string of cells which disappears below the caudal portion of the prosencephalon. During the subsequent development the differentiation of the notochord proceeds caudalwards and rostralwards. The notochord grows larger and terminates in the prechordal plate as shown by Adelmann (1926) and Spratt (1947, 1952). My observations in dissected and serially sectioned specimens confirm the findings of Mihalkovics (1877), Ahlborn (1883), Locy (1895) and Lehmann (1945), which showed that the notochord terminates at the boundary between the prosencephalon and the mesencephalon (Fig. 92). This boundary corresponds to the di-mesencephalic boundary of later ontogenetic stages. Accordingly the prosencephalon is prechordal, while the mesencephalon and the rhombencephalon (deuteroencephalon) are epichordal. Experimental investigations (*e.g.*, Nieuwkoop et al.,

1952, 1955; EYAL-GILADI, 1954; HARA, 1961) prove that the prechordal mesoderm induces the differentiation of the prosencephalon (situated above the prechordal plate) and the chordal mesoderm the mesencephalon, rhombencephalon and/or the spinal cord (situated above the notochord). These experiments fit with the abovementioned observations as substantiated also by my own findings on the relationship between the neural tube and the notochord. KINGSBURY (1924) investigated the development of the notochord and the neural tube in birds (Gallus) and fish (Squalus) and in addition performed experimental investigations in amphibians. He concluded that the notochord terminates ventral to the isthmic fovea at the boundary between the mesencephalon and the rhombencephalon. According to this view, the prosencephalon and mesencephalon are

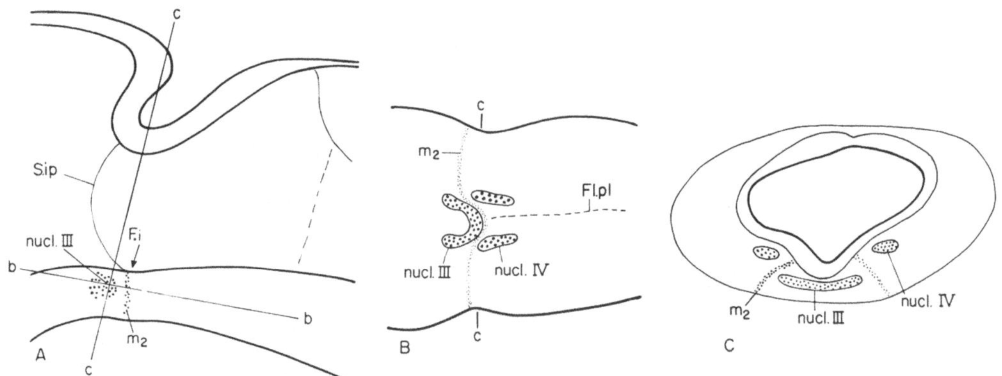

Fig. 80 A—C. Diagrams of parasagittal (A), frontal (B) and transverse (C) sections of the neural tube of chick embryo at stage *HH–27* showing the relationship between the caudal mesomere (m_2) and the isthmic segment; *b–b* and *c–c* indicate the level of sectioning of B and C, respectively

prechordal and the rhombencephalon epichordal. However, at stages *HH–11* and *HH–12* of chick embryo the notochord, as shown in Figs. 29 and 30 of KINGSBURY (1922), terminates between the second and the third bulge, *i.e.*, between *PrB* and m_1 as defined by me, and not between the mesencephalon and the rhombencephalon as interpreted by KINGSBURY. KINGSBURY described similar morphological pattern in fish (Squalus) and amphibians as in birds. But, the isthmic fovea (*F*) in his Figs. 25 and 26, illustrating two different ontogenetic stages of the neural tube in fish, are not identical according to my interpretations of the neural metamerism. In my opinion, KINGSBURY has misinterpreted the neuromeres in the mes-rhombencephalon and their relationship to the tip of the notochord. If so, no proof exists for the termination of the notochord at the isthmic fovea at early ontogenetic stages of the neural tube. Consequently neither a coextension between the notochord and the floor plate nor a neurochordal suture exist in the neural tube as supposed by KINGSBURY (1924).

b) The Isthmic Fovea and the Floor Plate

My observations confirm the findings of v. KUPFFER (1906), KINGSBURY (1920) and PALMGREN (1921), *viz.*, that the isthmic fovea develops rostralmost in the rhombencephalon (Fig. 80a). The fovea develops similarly in all vertebrates according to HIS (1892), v. KUPFFER (1906), HERRICK (1917), PALMGREN (1921) and KINGSBURY (1934). The isthmic fovea appears relatively late in the neurogenesis. All the same, the topographical relationship between the fovea and the remnants of the surrounding neuromeres, especially the m_2, can be accurately determined. The fovea may represent an important landmark of the rostralmost portion of the rhombencephalon. The floor plate and its derivative, the septum medullae, terminate rostrally in the rhombencephalon below the fovea isthmi (Fig. 80b) as maintained by KINGSBURY (1920, 1922, and 1934).

c) The Sulcus Intraencephalicus Posterior

The transformation of the ventricular sulcus of the caudal mesomere into the sulcus intraencephalicus posterior (Fig. 80) as described by v. KUPFFER (1906) and PALMGREN (1921) can be traced step by step. The sulcus and the cellfree mantle below it are conspicuous landmarks within the neural tube as discussed in chapter IV.

C. The Relationship between the Cranial Nerves (III—X) and the Neuromeres

The relationship between the neural crest and the cranial ganglions on the one hand and the neural tube on the other has been divergently described in the literature. According to MARSHALL (1878) the formation of the neural crest ("neural ridge") begins in the mesencephalic region in the chick of twenty-two hours incubation (*HH–8–9*) and spreads rostral- and caudalwards during the subsequent development in the forty three hours' chick (*HH–13*); the neural crest is recognizable from the summit of the optic vesicles and caudalwards to the postotic region. According to my observations, the neural crest is composed of three divisions located lateral to the pros-mesencephalon and the rh_2, the rh_4 and the *RhC*.

GORONOWITSCH (1893) divided the cranial neural crest into three portions called primary, secondary and tertiary ganglionic crests. The primary portion comprises the pros-mesencephalic region, and it arises before the secondary portion which includes the primordial trigeminal and the facial-acoustic ganglions. The tertiary neural crest is the last to appear. I have found in chick a corresponding developmental pattern represented by stages *HH–8*, *HH–9–10* and *HH–11*, respectively. However, I found that GORONOWITSCH's primary portion also embraces the primordial semilunar ganglion.

LILLIE (1919) described in chick the cranial neural crest as consisting of pre- and postotic portions which arise at different times. The preotic division is discernible at the 6—7 somite stage (*HH–9*). It extends from the extreme anterior end of the neural tube to about the centre of the auditory pit. The preotic crest looses its connection with the neural tube at stage *HH–10* (10 somites), and the cells mingle with the mesenchymal ones. The crest forms the rudiment of the semilunar and facial-acoustic ganglia. The postotic neural crest begins at the posterior margin of the otic pit at stage *HH–11* (13 s.) and extends caudalwards to the middle of the fourth somite. This portion of the crest is the primordial petro-nodose ganglion. According to my observations, LILLIE's preotic neural crest embraces both the pros-rhombencephalic and the preotic rhombencephalic divisions described by me. He does not mention, however, the absence of the crest formation in the rh_3 region. The pre- and postotic divisions are both present at stage *HH–10* and are interconnected by a thin string of cells in the roof of the neural tube. The connection is broken at stage *HH–13* where one crest portion is found lateral to rh_2, another lateral to rh_4 and a third lateral to *RhC*.

DI VIRGILIO et al. (1967) describe the neural crest as missing in the prosencephalon in birds. According to my observations in chick (Fig. 77), the neural crest extends uninterruptedly from the optic evagination along the roof of the mesencephalon and the rostralmost segment of the rhombencephalon (*RhA*).

DI VIRGILIO et al. state furthermore that the rhombencephalic portion of the neural crest makes its appearance shortly after the neural folds have fused. In view of my observations, this is true only prior to stage *HH–10*. After stage *HH–10–11*, the closure of the neural tube proceeds more rapidly than the growth of the neural crest. Therefore, the greater part of the crest develops from the dorsal wall of the cord itself and not from the angle between the ectoderm and the neural tube.

ADELMANN (1925) observed three neural crest proliferations in rat. The rostral part extends at the 4—5 somite embryo stage along the prospective mesencephalon and his rhombomere A_1, forming the rudiment of the semilunar ganglion. The preotic neural crest develops at the 7—8 somite embryo stage between the rh_3 and the otic pit, forming the rudiment of the facial-acoustic ganglion. The postotic neural crest follows shortly later (at the 8 somite stage) and forms the rudiment of the petro-nodose ganglion.

These three divisions of the cranial neural crest in rat seem to be homologous with those I have described in the chick. However, the preotic and postotic portions of the crest are interconnected in the chick, but according to ADELMANN (1925) and HOLMDAHL (1928) not in the rat.

The three portions of the crest can be traced step by step to their final differentiations into the semilunar, the facial-acoustic, the petrose and the nodose ganglions. These ganglions and their nerves are in chick related to the rhombomeres as previously described by HILL (1900), KAMON (1906) and NEAL (1918) among others. A different view is expressed by v. KUPFFER (1906), MEEK (1907), BERGQUIST and KÄLLÉN (1953a, 1954) as mentioned above (p. 46).

MEEK (1907), NEAL (1918) and PALMGREN (1921) hold that the oculomotor nucleus develops within the mesencephalon and the trochlear nucleus within the rhombencephalon, a conclusion to which my own observations lend support. The oculomotor and the trochlear nuclei are situated respectively rostral and caudal to mesomeie m_2, as shown in Fig. 80B. A different view is advocated by MCCLURE (1890), WATERS (1892), LOCY (1895), HILL (1900) and HUGOSSON (1957) who hold that the trochlear nucleus in the chick is situated within the mesencephalon. In my opinion this notion is based upon a misinterpretation of the mes-rhombencephalic boundary in the early ontogenesis, as the caudal mesomere is interposed between the oculomotor and the trochlear nuclei throughout the morphogenesis (Fig. 80).

The lateral half of the isthmic region, together with the trochlear nucleus, grows rostralwards relative to the ventromedial and dorsomedial regions of the neural tube. But the trochlear nucleus never migrates across the border into the mesencephalon. Nevertheless, the oculomotor and the trochlear nuclei may appear on the same transverse section through the brain stem (Fig. 80C), separated, however, by the caudal mesomere which is distinguishable until stages *HH–32—HH–34*. Bergquist and Källén (1954) locate the oculomotor nucleus within the caudal mesomere. This is not supported by me nor by Meek (1907) in chick.

When the abducent nerve has emerged from the ventral wall of the rh_3, the neuromeres m_1, Is, rh_2, rh_3, rh_4, rh_6 and rh_7 in chick are connected to the nerves III, IV, V, VI, VII and VIII, IX and X, respectively, as shown in Fig. 81. No cranial nerves emerge from the m_2, rh_1 and rh_5.

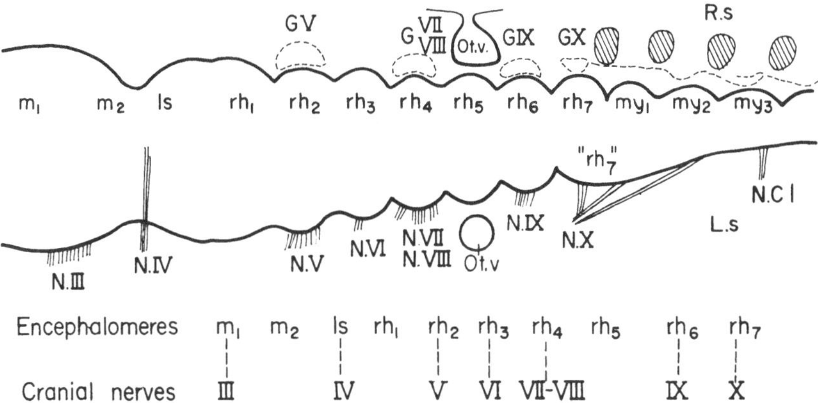

Fig. 81. Diagrammatic longitudinal section of the lateral wall of mesencephalons and rhombencephalons of chick embryos at stages *HH–13* (*R.s*) and *HH–24* (*L.s*). The anlage of the ganglions to the trigeminal (G_V), acustico-facial ($G_{VII-VIII}$), glossopharyngeal (G_{IX}) and the vagus (G_X), nerves are outlined on the right side. The oculomotor (*N III*), trochlear (*N V*), trigeminal (*N V*), abducens (*N VI*), acustico-facial (*N VII–VIII*), glossopharyngeal (*N IX*) and vagus (*N X*) nerves are outlined on the left side

Conclusion

In early developmental stages of the chick the neuromeres have a histological structure similar to the neuromeres defined by Orr.

The neuromeres occur both in the cerebral and the spinal divisions of the neural tube.

The notochord terminates rostrally at the pros-diencephalic boundary. The fovea isthmi develops rostrally in the rhombencephalon. The floor plate terminates at the mes-rhombencephalic boundary. The m_2-sulcus develops into the posterior intraencephalic sulcus marking the boundary between the mes- and rhombencephalon.

The oculomotor nerve emerges from the rostral mesomere, and the nerves IV, V, VI, VII—VIII, IX and X are connected with rhombomeres Is, rh_2, rh_3, rh_4, rh_6 and rh_7, respectively.

VII. Histochemical Studies on Sectioned Embryos

Sundberg (1924) described the occurrence of glycogen in the ventral raphe region of the epichordal brain. The correspondence between the distribution of the glycogen and the extent of the floor-plate was pointed out by Aloisi (1932) and Kingsbury (1934). But these investigators did not notice the characteristic correlation between the neuromeres and the distribution of glycogen. While studying the morphogenetic cell migration in the mes-rhombencephalon (Vaage, 1965) and the differentiation of the isthmic nuclei (Vaage, 1969) my attention was drawn to the question of the relation between neuromeres and glycogen, and also to a non-proliferating cell column located ventrally in the spinal cord and rhombencephalon from stage *HH–24* on. It was, therefore, deemed desirable to undertake a reinvestigation of the relationship of the non-proliferating cell column to the neuromeres, with particular reference to the occurrence of glycogen.

The proliferative potency of the neural epithelium is studied by autoradiographical investigation of the neural tube of (chick) embryos exposed to labelled thymidine for both a short (phase labelling, Fujita and Miyake, 1962) and a longer (cumulative labelling) time.

Thymidine is incorporated into desoxyribonucleic acid (DNA) of cells in the process of division (McQuade et al., 1955). The thymidine remains in these and their daughter cells (Hughes et al., 1958; Sidman et al., 1959a and b). The neural epithelium is heterogeneous and asynchronic, and its intermitotic period is about 10—14 hours at young stages (Fujita, 1962). In (chick) embryos with a well-developed vascular system (after $2^1/_2$—3 days incubation) injected thymidine is available to proliferating cells at least for 24 hours (Martin and Langman, 1965). Cells which do not incorporate labelled thymidine after prolonged exposure (24 hours) do not have proliferating activity during the actual period. If these cells are arranged in a column, it may be called a "non proliferating cell" column (*NPC*).

Results

PAS-Staining

Subjacent to the median longitudinal sulcus of the neural tube a PAS-positive substance occurs from stage *HH–20* on. Since the substance is absent following the preincubation in amylase, it evidently is glycogen (Pearce, 1960). The amount of glycogen increases considerably between stages *HH–20* and *HH–27* (Fig. 82) but remains restricted in its distribution. At the latter stage glycogen is found ventrally in the median raphe extending rostralwards to the rostral border (Fig. 82) of the isthmic fovea, but never passing beyond it. In other words the glycogen occurs ventrally in the spinal cord and the rhombencephalon, but is lacking in the floor of the pros- and mesencephalon (Fig. 92). The prospective trochlear nuclei (Fig. 83) are separated by the glycogen-containing raphe and consequently located within the *Is*-segment.

Autoradiography

At early developmental stages, 30 minutes after the injection of H^3-thymidine, the labelled cells are virtually restricted to the deeper half of the neural epithelium

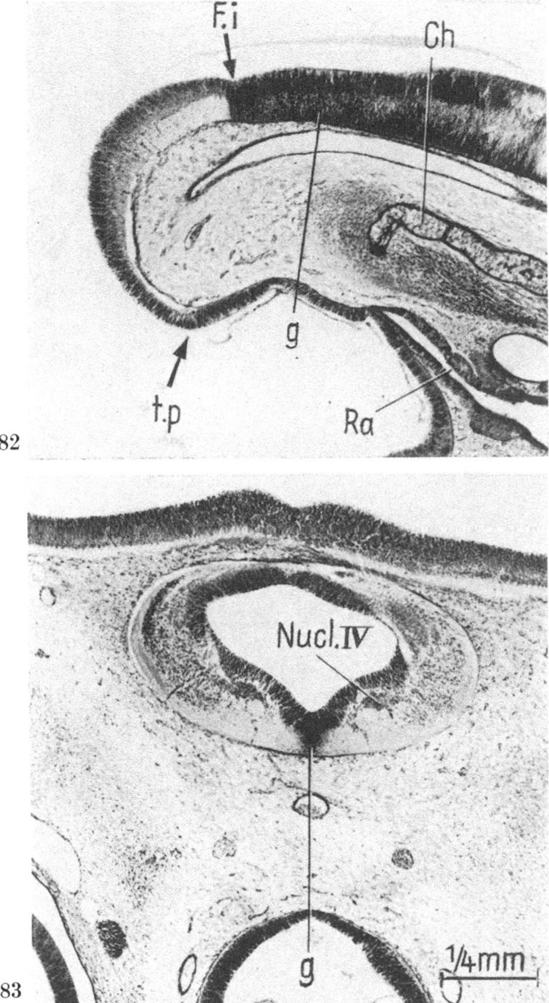

Figs. 82 and 83. Sagittal section (Fig. 82) through the mes-rhombencephalon and transverse section (Fig. 83) through the isthmic segment of chick embryo at stage *HH–27*, PAS-stained

around the whole neural tube. At stage *HH–24*, the neural epithelium is still heterogeneous and asynchronic in the pros- and mesencephalon (Fig. 84). Ventrally in the rhombencephalon (Fig. 86) and the spinal cord, however, there is an unlabelled column of cells (*NPC* Fig. 86). This column remains unlabelled even after a cumulative labelling. In control sections preincubated in pancreatic desoxyribonuclease no labelled cells occur (Vaage, 1969). The labelled thymidine is therefore incorporated in the DNA in cells which prepare for division. The aforementioned unlabelled cell column, which is coextensive with the distribution of glycogen, consists of non-proliferating cells. This column has an identical caudorostral extent at succeeding stages (*HH–27* and later) and terminates at the rostral border of the

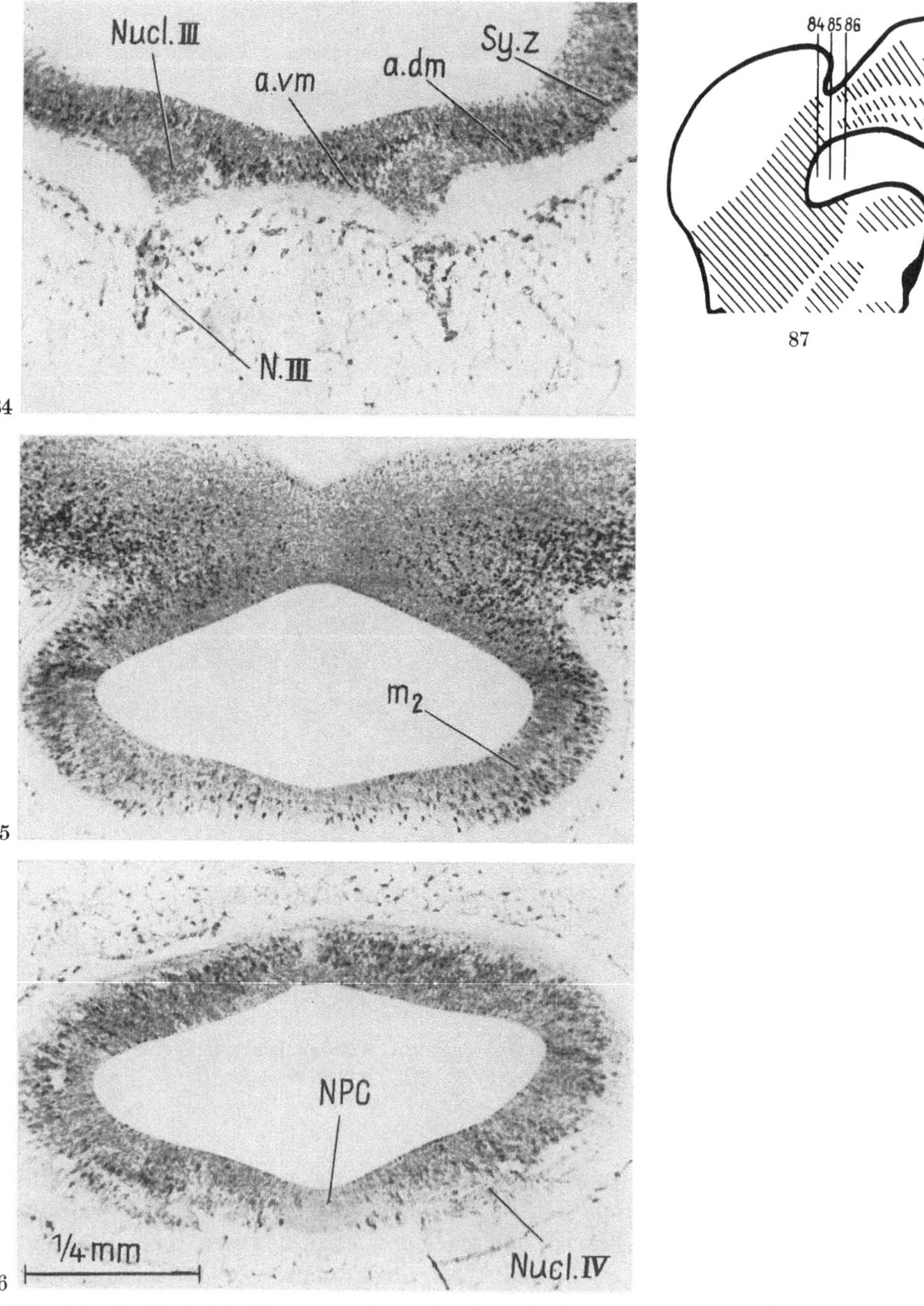

Figs. 84—86. Flash labelling autoradiographs through the m_1 (Fig. 84) m_2 (Fig. 85) and *Is* (Fig. 86) of chick embryo at the stage *HH-24*. The autoradiographs were taken 30 minutes after the injection of H^3-thymidine. Cells that are synthesizing DNA incorporate the labelled H^3-thymidine. Note that radioactivity is restricted to the cells in the synthetic zone at the mantle layer

Fig. 87. Diagrammatic lateral view of the cerebral tube of chick at stage *HH-24*. The hatched areas show incipient cell migration in the mantle layer. The transverse lines indicate the level of the sections with their respective numbers

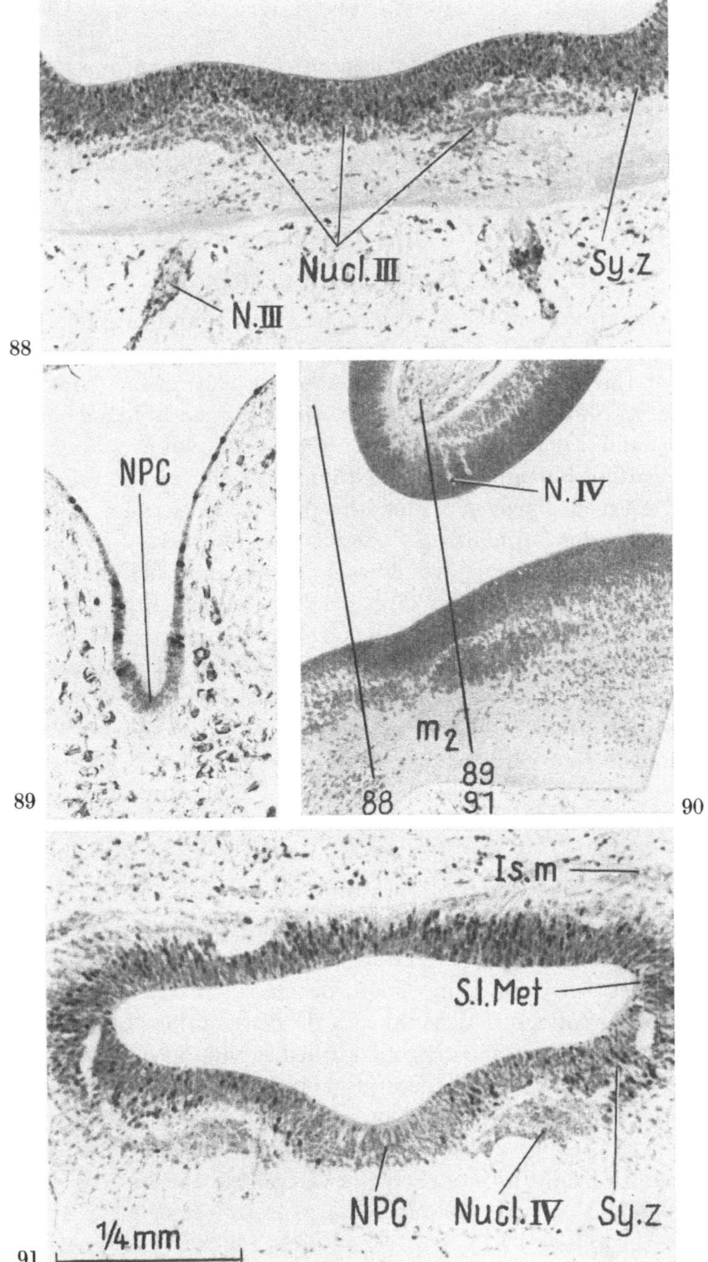

Figs. 88 and 91. Flash labelling autoradiographs through the m_1 (Fig. 88) and *Is* (Fig. 91) of chick embryo at stage *HH–27*. The autoradiographs were taken 30 minutes after the injection of H^3-thymidine. Cells that are synthesizing DNA incorporate the labelled H^3-thymidine. Labelled cells occur in the synthetic zone of the mantle layer. Note that no labelled cells (Fig. 91) are present in the septum rhombencephali

Fig. 89. Transverse section through the rostralmost end of isthmus of chick at stage *HH–46*. The chick was injected with H^3-thymidine at stage *HH–27*. Note that no labelled cells are present in the septum rhombencephali

Fig. 90. Parasagittal section through the mes-rhombencephalon of chick embryo at stage *HH–27*. The transverse lines indicate the level of the sections with their respective numbers

isthmic fovea. The trochlear nuclei are situated laterally to the rostralmost portion of the *NPC* (Fig. 91) while the oculomotor nuclei are located cranial to it (Figs. 88 and 90). The m_2 is interposed between the oculomotor and the trochlear nuclei (Fig. 90) just rostral to the tip of the *NPC*, which remains unlabelled even after a cumulative labelling (Fig. 89).

Discussion

A. The Glycogen Content

The glycogen appears in the ventral raphe of the neural tube of chick embryos prior to the formation of the isthmic fovea. At this early stage the caudorostral distribution of the glycogen accumulation ceases rostrally in the isthmus just caudal to the m_2. From stage *HH–24* on, the glycogen is found ventral to the isthmic fovea and caudalwards (Aloisi, 1932; Kingsbury, 1934). Sundberg (1924), investigating human embryos of 15 mm, 27 mm and 40 mm CrR lengths, consistently observed glycogen within the floor-plate (septum medullae) ,,vom Mesencephalon bis zum caudalsten Ende des Rückenmarks''. In chick embryos, the glycogen-containing raphe never crosses the boundary between the m_2 and *Is*. The notochord likewise is rich in glycogen. Sundberg (1924) and Kingsbury (1934) mentioned that the notochord and the glycogen-containing raphe are co-extensive. As shown above, however, the notochord in chick terminates rostrally in the mammillary region. In other words, the notochord and the glycogen-containing ventral raphe are not co-extensive either at stage *HH–24*, or at later stages. Consequently, no correlation seems to exist between the notochord and the glycogen-containing ventral raphe as supposed by Sundberg (1924) and Kingsbury (1934).

B. The Non-Proliferating Cell Column (NPC)

The *NPC* occurs in chick embryos from the 3 day stage on, while the neuromeres are still discernible. The column terminates ventral to the isthmic fovea at the boundary between mes- and rhombencephalon. Consequently the caudorostral extent of the column is identical with the distribution of the raphe glycogen (Fig. 92). Cells which do not incorporate labelled thymidine even after cumulative (24 h) labelling have lost the proliferating activity (Fujita and Miyake, 1962). Kingsbury (1922) described the floor plate in chick as differentiating into ependymal cells in the 3 to 4 day chick, concomitantly with the differentiation of the *NPC* as shown by the author. Some of the cells in the *NPC* at stage *HH–27* are presumably identical with the cells in Kingsbury's floor-plate. It is probable, therefore, that the author's *NPC* is an expression of the early differentiation of the neural epithelium into ependymal cells ventrally in the neural tube. Coghill (1924) described both ependymal cells and neuroblasts within the floor-plate of amphibians. It is still necessary to investigate whether the non-proliferating cell column in chick consists of both neuronal and non-neuronal elements at early developmental stages.

Conclusion

The glycogen occurs within the floor-plate from the third day on, in an area which terminates rostrally just caudal to the mes-rhombencephalic boundary.

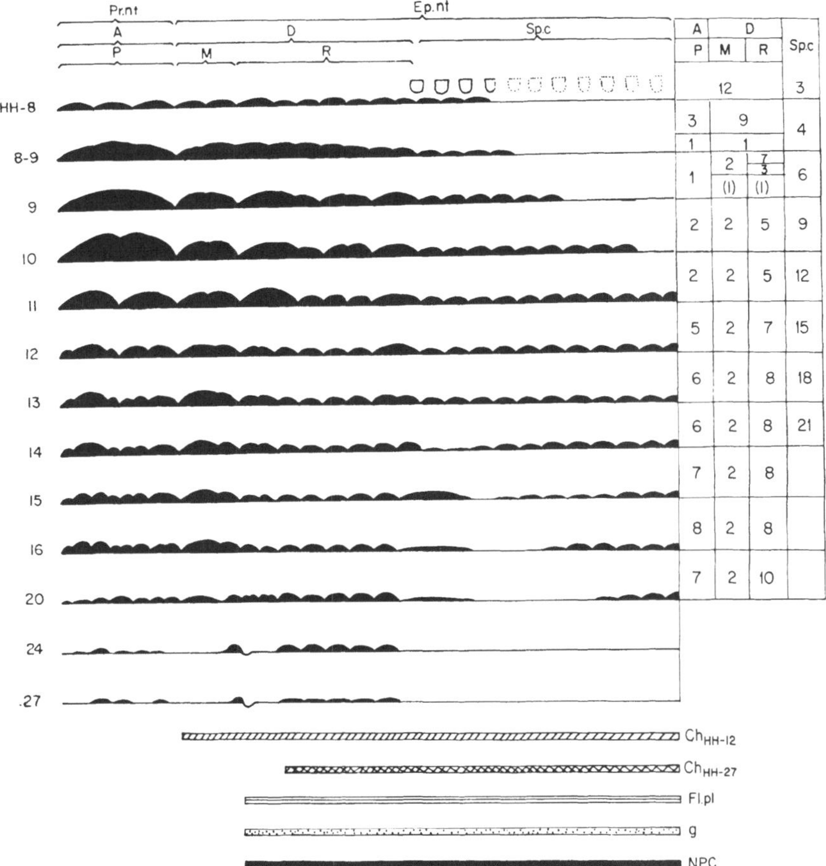

Fig. 92. Diagrams showing the degree of neuromery and the number of neuromeres at different stages of development of chick embryos. The black bulges mark the neuromeres. The subdivision of the neural tube into prechordal (*Pr.c.*), epichordal (*Ep.c.*), archencephalic (*A*), deuteroencephalic (*D*), spinal cord (*Sp.C.*), pros- (*P*), mes- (*M*) and rhombencephalic (*R*) regions is outlined above the figure. The extent of the notochord (*ch*), the floor plate (*Fl.pl*), the content of glycogen (*g*) and the NPC-column (*NPC*) are indicated below the figure. The stages are listed on the left side and the number of neuromeres in arch-deuteroencephalon, pros-, mes- and rhombencephalon and the spinal cord at each particular stage are listed on the right side of the drawing

The non-proliferating cell column, like the floor plate and the distribution of glycogen in the raphe, terminates rostrally in the *Is*-segment at early stages and below the isthmic fovea at later stages of development.

VIII. General Discussion

The extensive literature dealing with the early development of the nervous system (*e.g.*, MIHALKOVICS, 1887; v. KUPFFER, 1906; MEEK, 1910; BERGQUIST and KÄLLÉN, 1954) presents strong evidence supporting the idea that the morphogenesis of the central nervous system is in principle identical in all vertebrates. Hence the observations in chick reported in the present study should form a good foundation for discussion of the conflicting views with regard to the early morphogenesis of the central nervous system of vertebrates in general.

It seems appropriate to start with the question of the neuromeres. Do the neuromeres exist and, if so, do they occur throughout the neural tube? My observations in the chick warrant affirmative answers to both these questions. Living specimens as well as fixed preparations give evidence of a progressive differentiation of the neural tube into bulges displaying the characteristics of neuromeres as defined by ORR (1887). NEAL (1918) and STREETER (1933), pointing out the developmental coincidence between the myelomeres and the somites, held that the myelomeres are not segments of the neural tube comparable with the rhombomeres but are caused by pressure from the somites. My observations in the chick, however, show that the myelomeres and the encephalomeres at early stages of development have similar histological structures corresponding to that of the neuromeres as defined by ORR (1887). The evidence, therefore, definitely favours the conclusion that the entire neural tube is differentiated into neuromeres during the early morphogenesis.

Acceptance of the doctrine of neuromerism prompts the question of the number of neuromeres. As shown diagrammatically in Figs. 40 and 92, the number of neuromeres in chick varies with the stage of development. New encephalomeres are formed by division of existing encephalomeres, though new myelomeres always are formed by progressive differentiation of the spinal tube in caudal direction. A similar morphogenesis of the neural tube in different vertebrates is repeatedly described in the literature, as shown in table 1. The pattern presumably holds for all vertebrates.

As described above, examinations of closely graded stages of chick embryos clearly show that successive encephalomeres are formed by division of the preceeding ones. Consequently, the number of neuromeres is dependent on the stage of development. The discrepancies found in the literature with regard to the number of neuromeres may be readily explained by assuming that the various authors have based their conclusions on examinations of different developmental stages. If this assumption is correct, the neuromeres described by LOCY (1895) are identical with PrA_1, pr_{4-6}, PrB_2, m_{1-2}, RhA_1, rh_{2-5}, RhC; those of NEAL correspond to P, M, RhA_1, rh_{3-5}, RhC; and those described by NEUMAYER are identical with PrA, PrB_1, PrB_2, m_{1-2}, rh_{1-7}. The prosomeres described by RENDAHL (1924) correspond to pr_{1-2}, pr_{3-5}, pr_6 and prB_2; and those of KUHLENBECK (1954) correspond to pr_{1-2}, pr_3–pr_6, PrB_2.

Since DOHRN'S (1875) theory of an annilid origin for the vertebrates, morphologists have stated that the vertebrate head exhibits a segmented morphology. If so, each segment must embody a neuromere, a head somite, or a branchial anlage, and a nerve (KINGSBURY, 1926).

This phylogenetic interpretation of the vertebrate head results in a numerical correspondence between the neuromeres, the nerves and the mesodermal segments. Thus the number of neuromeres should correspond to the number of somites and nerves. Does such a correspondence exist? My observations in the chick confirm that the metamerism in the trunk region corresponds to the neuromerism within the spinal tube. The myelomeres develop concomitantly with the somites, and a numerical correspondence exists. One portion of the neural crest and one spinal nerve correspond to each myelomere as advocated repeatedly in the literature (McClure, 1890; Minot, 1892; Johnston, 1916; Neal, 1918). A correspondence exists between the spinal neuromerism and the trunk metamerism.

In the head region the relation between the neuromeres, the nerves and the mesodermal segments is still open to question. Much effort has been devoted to reveal the metameric structure of the vertebrate head similar to that in the annilid (Dohrn, 1875), but the results of the researches have not yet led to unanimity of opinion. In lower vertebrates (Elasmobranchs) the head mesoderm is segmented into somites reaching forward nearly as far as the anterior end of the notochord. By far the greatest part of the branchial region is occupied by the somites (Gegenbaur, 1887; Froriep, 1902; Ziegler, 1908; Goodrich, 1918; Veit, 1939). In general the number and arrangement of the pharyngeal arches and grooves corresponds to that of the somites. The location of the grooves and slits is intersegmental, each having broken through between two mesodermal somites. According to Ziegler (1908) and Goodrich (1918) among others, there is a coincidence between the branchiomerism and the metamerism in vertebrates. In higher vertebrates (amphibians and amniotes) the mesoderm in front of the vagus is not divided into metameric segments (Ahlborn, 1883; Froriep, 1887; Starck, 1943). The piercing of the pharyngeal slits makes the mesoderm appear segmented, producing the branchiomerism. The latter, taking its origin from the entoderm, has nothing to do with the metamerism which takes its origin from the mesoderm (Ahlborn, 1883; Starck, 1943). According to Starck, the branchiomerism and the metamerism do not correspond to each other in any vertebrates. However, both the branchiomeres and the metameres are formed by serial repetition of structures (Versluys, 1923; Adelmann, 1926; Kingsbury, 1926). The pharyngeal arches vary in different vertebrates and show individual variations (Cyclostoma) and variation during the life span (Starck, 1943). Thus Petromyzon has eight pharyngeal arches, Squalus seven, Necturus six, chick five and rabbit four (Kingsbury, 1926). The encephalomeres do not correspond with the variations in the mesodermal head segments but differ from them with respect to their mode of formation, their number and their homology. 1. The encephalomeres are formed by repeated subdivision of already existing encephalomeres (Figs. 39, 40 and 92), resulting in the formation of smaller and smaller bulges. This mode of differentiation contradicts the view advocated by Källén (1954) that the encephalomeres are serially repeated structures occuring along the axis of the neural tube. According to the definition of segmentation given by Rugh (1943), the encephalomeres are not true segments comparable with the mesodermal ones which are serially repeated embryonic rudiments in successive levels of the axis of embryo. 2. A corresponding number of encephalomeres exist throughout the vertebrate range (see below), while the pharyngeal arches and head somites vary (see above). It is evident that there is no constant relationship between the neuromerism and the branchiomerism and metamerism in the head region. 3. The encephalomeres are individually homologous structures in different vertebrates, as maintained repeatedly in the literature (Meek, 1907, 1910; Bergquist and Källén, 1954), while the mesodermal segments are serially homologous (Kingsbury, 1926). In my opinion these differences between the encephalomeres and the mesodermal head segments militate against the theory of numerical correspondence between encephalomeres and mesodermal head segments, as proposed by *a.o.*, Ziegler (1908), Goodrich (1918) and Delsman (1922). In the head region the neuromerism does not correspond numerically with the metamerism and the branchiomerism in any vertebrates as far as can be seen from the literature. Hence the phylogenetic interpretation held by Delsman (1922) that the neuromerism is secondary to the metamerism and branchiomerism must be rejected.

The cranial nerves (III—X) do not correspond numerically to the metamerism, the branchiomerism and the neuromerism in the head region, but they appear to bear a fixed relation to the encephalomeres (Fig. 81) throughout the vertebrate range. Thus, the number of encephalomeres cannot be determined by study of the head morphology or the number of cranial nerves. But the remarkable constancy of the number of encephalomeres in different vertebrates strongly emphasizes their character as morphological entities within the neural tube. As mentioned above, differences in the numbers of encephalomeres found by various investigators evidently result from examinations of embryos in different stages of development.

Let us now turn to the factors causing formation of the encephalomeres. It has already been mentioned that head segmentation, cranial nerves and neural crest division are of little importance in this respect. Elimination of these extrinsic factors prompts the question of possible intrinsic causes. Recent embryological experiments have shed new light on this question. Nieuwkoop et al. (1952) defined the activating and transforming principles and showed that they induce the prospective neural plate, which then acquires potencies for autonomous differentiation (Eyal-Giladi, 1954; Boterenbrood, 1962; Nieuwkoop, 1965). Inductive actions of short duration lead to the formation of a segmentation pattern closely resembling the normal pattern but differing from it in certain respects (Nieuwkoop, 1965). The occurrence of a normal pattern of encephalomeres depends upon the presence of surrounding mesoderm, especially underlying notochordal tissue (Källén, 1956c). This proves that the presumptive neural plate already contains all the factors necessary for further development, as well as ultimate cellular differentiation, but that inductive influences from the surrounding mesoderm are co-responsible for the establishment of the normal segmentation pattern. Adelmann (1936) maintains that the inductive influences responsible for the establishment of the normal pattern and the normal quantitative composition of the neural tube are of inhibitory character and emanate from the median portion of the underlying archenteron roof. The inductive factors described by Källén and Adelmann may be identical. Do the induction experiments shed light on the causes of the normal segmentational pattern of the neural tube? In my opinion the answer is affirmative. As mentioned above, inductive actions of the activating and the transforming principles (Fig. 40) convert prospective ectoderm into neural tissue. Varying intensity of the transforming principle on the presumptive neural tissue appears to be responsible for the regional organization of the rudimentary nervous system (Nieuwkoop, 1965). Further growth is governed by intrinsic autonomous factors and inductive stimuli from the surrounding mesoderm. These inductive stimuli may be derived from localized induction centres possibly comparable with the head and the trunk organizer of Speeman (1918) at early neurogenesis, and with the archencephalic, the deuteroencephalic and the spinocaudal inductive zones of Lehmann (1945) at later stages of neurogenesis. The regional correspondence between the induction centres and the divisions of the neural tube support the abovementioned assumption (Fig. 40). Thus the cerebral tube, which is the centre for formation of the neural plate segments of Locy (1895) and Hill (1899, 1900) and of primary encephalomeres, and the spinal tube correspond to the head and the trunk organizer of Speeman (1918) at early neurogenesis. At later stages the formation of neuromeres proceeds simultaneously and in different ways within the prosencephalon, the rhombencephalon and the spinal tube, and the three divisions of the neural tube correspond to Lehmann's three induction centres.

The supposed existence of a rostro-caudally spreading organizing factor within the central nervous system is unsatisfactory as an explanation of the neuromerism (Källén, 1956c), since the cerebral tube does not differentiate rostrocaudally but rather caudo-rostrally at early stages of development, as previously advocated by Locy (1895), Neal (1918), Bartelmez (1923), Lehmann (1945) and Nieuwkoop et al. (1955). Bartelmez states: "In the series at our disposal the division of the primary segments in general proceeds forward from the hindbrain" and "it seems probably that in man the differentiation of secondary neuromere usually begins caudally and proceeds forward" (*l.c.*, p. 244). His Fig. 6 illustrates this caudo-rostral differentiation.

The problem of the pattern of the disappearance of the neuromeres is closely related to the pattern of the morphogenesis of the neural tube. The reduction of the encephalomeres occurs gradually and concomitantly in all three brain regions. Some of the neuromeric sulci do not disappear, and all of the myelomeres undergo reduction in a rostrocaudal direction in accordance with the statements of among others Minot (1892). The two waves of reduction described by Bergquist and Källén (1954) are not observed.

Those neuromeric sulci which remain are transformed into ventricular sulci. But do the neuromeric sulci persist permanently during neurogenesis? In view of my observations in chick the answer is affirmative. The development of the neuromeric sulci into permanent ventricular sulci can be traced from stage to stage, in accordance with the findings of Locy (1895), v. Kupffer (1906), Palmgren (1921) and Herrick (1948). Thus the pr_3-sulcus can be traced into the sulcus intraencephalicus anterior of v. Kupffer, the "pr_{4-5}"-sulcus into the medial and ventral thalamic sulcus of Haller (1929), the pr_7-sulcus into the metathalamic recess of Haller and the m_2-sulcus into the posterior intraencephalic sulcus of v. Kupffer. Bergquist and Källén (1954) state that the ventricular sulci are proliferation furrows migrating on the wall of the neural tube. Consequently the sulci cannot be used as a point of referance during the morphogenesis of the neural tube. Bergquist and Källén illustrate a migration of prosomeric sulci at early neurogenesis. Källén thus describes a furrow (*S.i.a*) which "connects the first signs of a hemispheric evagination with the optic evagination. It seems to be homologous to the reptilian sulcus i.a." (Källén, 1951a, p. 7). In subsequent stages of human embryos (6,5 mm and 10,5 mm) the "furrow runs in the ventricular wall from the optic evagination up to the caudal border of the foramen Monroi" (*l.c.*, p. 7). According to my observations in the chick, the former sulcus is identical with the PrA_1-sulcus and the other with the pr_3-sulcus present at two different stages of development. BERGQUIST (1952a) observes that the sulcus within neuromere I extends into the optic stalk at early stages of development of both fish (Petromyzon) and reptiles (Lacerta). At later stages of development the sulcus within neuromere II extends into the stalk. He interpretes this as a migration of the sulcus from one neuromere into another. Judging from the morphological pattern found in the chick, BERGQUIST's neuromere I at early stages of development corresponds to PrA_1, while his neuromere II at later stages corresponds to pr_3. Hence the two sulci apparently are situated within two non-identical prosomeres. If this interpretation is correct, the results of BERGQUIST and KÄLLÉN do not militate against the conclusion that a relationship exists between neuromeric sulci and certain permanent brain sulci.

KUHLENBECK (1954) looked upon the neuromeres as temporary structures, the furrows of which are transformed into longitudinal sulci. As far as I can see, the encephalomeric sulci are transverse, while the longitudinal sulci within the prosmesencephalon cross several encephalomeres.

The question of the morphological subdivision of the brain, to be considered next, is closely related to the problem of the neuromeres and their boundaries.

Study of the early morphogenesis of the central nervous system of the chick lends support to the view advocated *e.g.*, by NEUMAYER (1899), HILL (1900), ZIEHEN (1906) and HOCHSTETTER (1919). The embryological evidence favours the distinction of four major cerebral subdivisions: telencephalon, diencephalon, mesencephalon and rhombencephalon. The definition of the boundaries between these subdivisions and the boundary between the rhombencephalon and the spinal cord has been and still is a controversial and much discussed matter.

The Tel-Diencephalic Boundary is an intraprosencephalic border interposed between prosomeres. The boundary between the *PrA* and *PrB* is the first to be formed during the subdivision of the prosencephalon (Fig. 29), and an identical boundary presumably occurs in different vertebrates (HIS, 1888; NEUMAYER, 1899; ZIEHEN, 1906). This boundary ($F.t\text{–}d_1$, Fig. 25) is the most natural tel-diencephalic border, as emphasized also by the Anatomische Gesellschaft (according to ZIEHEN, 1906, p. 279). LOCY (1895), MEEK (1907) and GRÜNTHAL (1952) claim that the border in question extends to the $F.t\text{–}d_2$ in Fig. 25, and v. KUPFFER held that the anterior intraencephalic sulcus (identical with the pr_3-sulcus) marks the boundary in question. In view of my observations the disagreement is a question of definition of the boundary. The $F.t\text{–}d_2$ and the pr_3-sulcus develop at a later developmental stage than the $F.t\text{–}d_1$, and I cannot see that they are more likely boundaries than the $F.t\text{–}d_1$. The tel-diencephalic boundary is a morphological border between temporary neuromeres as are the boundaries between the other brain regions at a corresponding developmental stage. BERGQUIST and KÄLLÉN (1954) reject the existence of a tel-diencephalic boundary of morphological significance. In my opinion, the boundary exists for a short time, subsequently disappearing concomitantly with the reduction of the prosomeres.

The Di-Mesencephalic Boundary was placed immediately caudal to the posterior commissure by NEAL (1898) and this view is supported by my observations in chick (p. 33). However, HERRICK (1948) in amphibians and apparently KUHLENBECK (1954) in all vertebrates put the border rostral to the posterior commissure.

My observations in the chick show that this boundary is identical with the interprosomeric border between pr_7 and pr_8 within the PrB_2.

The Mes-Rhombencephalic Boundary. At early stages of development the mesrhombencephalic fissure marks the boundary between the mes- and rhombencephalon. In the chick it is evident that the m_2-segment undergoes reduction and persists as the posterior intraencephalic sulcus which marks the mes-rhombencephalic boundary. The reduced m_2-segment is very conspicuous during the early neurogenesis in all vertebrates examined (PALMGREN, 1921). The boundary passes immediately caudal to the oculomotor nucleus which consequently belongs to the m_1, and not rostral to it as interpreted by AHLBORN (1883). KINGSBURY (1922, 1934) pointed out that the rostral end of the floor plate, the isthmic fovea and the glycogen-containing raphe can be used as fixed references for this boundary. His observations are supported by my findings in chick.

The Rhombo-Spinal Boundary is medial to the first somite in a variety of vertebrates according to MALPIGHI (1673), v. BAER (1828), MIHALKOVICS (1877), McCLURE (1890), MINOT (1892), LOCY (1895), BOMAN (1896), HILL (1900), BARTELMEZ (1923) and others. However, a number of embryologists (ZIMMERMANN, 1892; MINOT, 1892; MEEK, 1907, 1909, 1910; JOHNSTON, 1916; BERGQUIST, 1952a) adopted the (secondary) border between the definite spinal cord and the myelencephalon as the rhombo-spinal boundary.

Thus JOHNSTON states: "Since the first cervical nerve belongs to the tenth neuromere, the preceding nine belong to the rhombencephalon. The lower limit of the rhombencephalon may, therefore, be definitely defined in young stages as the constriction between the ninth and tenth neuromeres. This lies opposite the last pair of occipital somites" (*l.c.*, p. 210). In the chick, the rhombo-spinal boundary is transitory and divides the neural tube into a cerebral and a spinal tube. The border vanishes concomitantly with the enlargement of the caudal rhombomere, formed by the last rhombomere and the rostralmost myelomeres. After this transformation, a secondary rhombencephalon (myelencephalon) is formed by the rhombomeres and the adjacent myelomeres. A corresponding development presumably occurs in reptiles, birds and mammals (see above). The myelo-spinal boundary develops in amniota, parallel with the evolution of the spinal accessory and the hypoglossal nerves according to MINOT (1892). In my opinion it is appropriate to reserve the designation rhombo-spinal boundary for the embryological boundary between the primary rhombencephalon (*RhA–RhC*) and the spinal cord (as stated on p. 34). The myelo-spinal boundary, on the other hand, should be employed for the border between the secondary rhombencephalon and the spinal tube in amniota.

As mentioned above, there is no numerical correspondence between the cranial nerves (III—X) and the neuromeres. But the nerves bear a constant relation to certain neuromeres throughout the vertebrate range, while other neuromeres are devoid of peripheral nerves. In chick the rh_2, rh_3, rh_4, rh_6 and rh_7 are connected with the nerves V, VI, VII—VIII, IX and X, respectively (Fig. 81), while the neuromeres rh_1 and rh_5 have no such direct peripheral relation as shown by ORR (1887), HOFFMANN (1889), PLATT (1891), ZIMMERMANN (1891), WATERS (1892), BROMAN (1896), NEUMAYER (1899 and 1900), HILL (1900), BRADLEY (1904, 1906),

INGALL (1906), KAMON (1906), ZIEHEN (1906), THOMPSON (1907), HOCHSTETTER (1919), BARTELMEZ (1923), ADELMANN (1925) and SCHUMACHER (1928).

v. KUPFFER (1906) and BERGQUIST and KÄLLÉN (1953a, 1954) on the other hand, found that only the equivalent of the rh_1 lacks direct peripheral connection, while MEEK (1909, 1910) maintained a similar view concerning the equivalents of neuromeres rh_5 and rh_6 as well.

As far as I can see, the discrepancies are due to misinterpretation of some rhombomeres (cf. p. 46 and p. 66).

The eye-muscle nerves III and IV present a special problem with respect to their relationship to the mes- and rhombencephalon. My findings in the chick show that the oculomotor nucleus differentiates and remains within the floor of the mesencephalon, while the trochlear nucleus is situated in the floor of the rhombencephalon, thus supporting previous findings by MIHALKOVICS (1877), HIS (1888), HOFFMANN (1889), BRADLEY (1906), ZIEHEN (1906), THOMPSON (1907), NEAL (1918), HOCHSTETTER (1919) and PALMGREN (1921) in various vertebrates.

NEAL (1918) pointed out that the trochlear nucleus (in mus musculus) migrates from the rostralmost portion of the rhombencephalon into the mesencephalon, and BENGMARK et al. (1953) supported this view. Before its migration „Der Trochleariskern liegt in diesem Stadium lateral und etwas caudal von dem unpaaren, medianen Oculomotoriuskern“ (BENGMARK et al., p. 80). Still others, *e.g.*, MCCLURE (1890), WATERS (1892), HILL (1900) and HUGOSSON (1957), assert that the trochlear nucleus differentiates and remains within the floor of the mesencephalon in all vertebrates. AHLBORN (1883), however, held that both the oculomotor and the trochlear nuclei develop and persist within the floor of the rhombencephalon.

As described above (p. 66), in the chick the mantle of the caudal mesomere is interposed between the two nuclei in question throughout the course of histogenesis. BENGMARK et al. outline on their graphical reconstructions (Figs. 2, 10—12) a discontinuity in the longitudinal cell columns within the floor of the mesencephalon. The relationship of the discontinuity to the third and fourth nucleus is similar to the relationship between the m_2 and the oculomotor and trochlear nucleus in birds. The discontinuity of BENGMARK et al. is therefore likely identical with my m_2. The conclusions of BENGMARK et al., therefore, most likely are due to a misinterpretation of the mes-rhombencephalic boundary at certain stages of development.

The location of the mes-rhombencephalic border has been a matter of discussion for a long time. Several secondary structures (*e.g.*, the notochord, the floor plate and the isthmic fovea) are used conventionally without any attempt to define their relationship to the interneuromeric boundaries. The structures develop during the neuromeric period but their relations to the neuromeres have not been defined to my knowledge.

My findings in chick show that *the notochord* terminates below the PrB_2 near the di-mesencephalic boundary, in conformity with the findings of v. BAER (1828), FOSTER and BALFOUR (1876), AHLBORN (1883), HILLEMANN (1943) and PATTEN (1958). *The floor plate* extends rostrally to the mes-rhombencephalic boundary just caudal to the m_2-segment, as found by KINGSBURY (1922, 1934). AHLBORN (1883) and KINGSBURY (1922, 1924, 1934) described a correspondence between the notochord and the floor plate in different vertebrates. In view of my findings in the chick, the two structures in question are not coextensive, but terminate at different levels of the tube as mentioned above. *The glycogen* occurs within the floor plate prior to its ultimate differentiation, and its extent can be accurately determined. The non-proliferating cell column is coextensive with the glycogen-containing raphe, is present from the late neuromeric period in the chick.

The Isthmic Fovea develops rostrally in the *Is*-segment in the chick and presumably in all vertebrates, as discussed in Chapter VI.

The Posterior Intraencephalic Sulcus lies within the m_2-segment in all vertebrates (see Chapter VI). Below the sulcus the mantle layer is cellfree, marking the mes-rhombencephalic boundary. Thus both the posterior intraencephalic sulcus and cellfree mantle layer of m_2 are conspicuous morphological landmarks. Ependymal elements form a dense band of parallel fibres which originate in the ventricular sulcus, run through the mantle layer, and terminate at the pial surface according to HERRICK (1948). HERRICK also states that the fibres can be demonstrated by means of the Golgi procedure. Thus the relationship between the secondary structures mentioned and the neural tube as shown in Fig. 92 and discussed in Chapter VI and VII are identical in all vertebrates.

Summary

The segmentation of the primitive neural tube in chick is investigated in living and fixed chick embryos and the following conclusions are made:

A. The patterns of segmentation are different in the cerebral tube and the spinal tube.

a) At early ontogenetic stages the encephalomeres and the myelomeres have a histological structure conforming to that of the neuromeres of ORR.

b) The cerebral tube differentiates initially in a caudorostral direction, the spinal tube in a rostrocaudal direction.

c) At later stages segmentation proceeds differently in the prosencephalon, the rhombencephalon and the spinal tube.

d) Homologous neuromeres presumably occur in all vertebrates.

B. A topographical relationship exists between the cranial nerves and the neuromeres. A similar relationship presumably exists in other vertebrates.

C. The intracerebral boundaries develop from interneuromeric or neuromeric structures present at early developmental stages.

a) The first interprosomeric boundary to develop extends caudal to the infundibular region, the second extends rostral to the optic nerve.

b) The pros-mesencephalic boundary is located caudal to the posterior commissure.

c) The mes-rhombencephalic boundary runs between the oculomotor and the trochlear nucleus.

d) The rhombo-spinal boundary lies at the level of the rostralmost somite.

D. The notochord extends rostralwards to the boundary between di- and mesencephalon. The floor plate, the raphe-containing glycogen and the non-proliferating cell column terminate just caudal to the mes-rhombencephalic boundary. The isthmic fovea occurs rostrally in the rhombencephalon. The m_2-sulcus turns into the boundary sulcus between the mes- and rhombencephalon.

Acknowledgements. The author is deeply indebted to Professor JAN JANSEN, M.D. for his encouragement and never-failing interest in this study and for his advice and invaluable criticism during the preparation of the manuscript. It is a great pleasure to acknowledge my profound indebtedness to Professor WILHELM HARKMARK, M.D. for stimulating discussions and critical advice. I also want to express my sincere gratitude to Mrs. BERIT BRANIL, Mrs. LIV DALE, Mrs. INGER GRØHOLT, Mrs. AGNES HOLTER, Mrs. EVELYN PETTERSEN and Mr. EINAR RISNES for skilful assistance and to my colleagues at the Anatomical Institute for their interest and support.

Bibliography

ADELMANN, H. B.: The development of the neural folds and cranial ganglia of the rat. J. comp. Neurol. **39**, 19—171 (1925).

— The development of the premandibular head cavities and the relations of the anterior end of the notochord in the chick and robin. J. Morph. **42**, 371—439 (1926).

— The problem of cyclopia. I and II. Quart. Rev. Biol. **2**, 161—182, 284—304 (1936).

AHLBORN, F.: Untersuchungen über das Gehirn der Petromyzonten. Z. wiss. Zool. **39**, 191—294 (1883).

ALOISI, M.: La Sede ed il limite rostrale del Glicogeno nel Nevrasse in Mammiferi durante lo sviluppo. Boll. Soc. ital. Biol. sper. **7**, 1—4 (1932).

AREY, L. B.: Developmental anatomy. London: W. B. Saunders Company 1965.

BAER, K. E. v.: Über die Entwicklungsgeschichte der Thiere, Bd. 1—3. Königsberg 1828.

BARTELMEZ, G. W.: The subdivisions of the neural folds in man. J. comp. Neurol. **35**, 231—247 (1923).

—, and H. M. EVANS: The development of the human embryo during the period of somite formation including embryos with 2 to 16 pairs of somites. Contr. Embryol. Carneg. Inst **17**, 1—67 (1926).

BEER, G. R., and E. J. W. BERRINGTON: The segmentation and chondrification of the skull of the duck. Phil. Trans. B **223**, 411—468 (1934).

BENGMARK, S., R. HUGOSSON u. B. KÄLLÉN: Studien über Kernanlagen im Mesencephalon sowie im Rostralteil des Rhombencephalon von Mus musculus. Z. Anat. Entwickl.-Gesch. **117**, 73—91 (1953).

BERANECK, E.: Etude sur les replis medullaires du poulet. Recu. zool. Suisse **4**, 305—364 (1887).

BERGQUIST, H.: Zur Morphologie des Zwischenhirns bei niederen Wirbeltieren. Acta zool. (Stockh.) **13**, 57—303 (1932).

— Studies on the cerebral tube in vertebrates. The neuromeres. Acta zool. (Stockh.) **33**, 117—187 (1952a).

— Transversal bands and migration areas in Lepidochelys olivacea. Acta Univ. lund. Avd. 2, **63**, Nr 13, 19 p. (1952b).

— Die Neuromerie. Anat. Anz. **102**, 449—456 (1956).

— Mitotic activity during successive migration in the diencephalon of chick embryos. Experientia (Basel) **13**, 1—6 (1957a).

— Comments on the architype of the vertebrate brain. Kungl. Fysiogr. Sällsk. Lund Förh. **27**, 153—160 (1957b).

— Mitotic activity in early phases of CNS-development. Proceedings of a symposium held in Prague on November 12th—13th, 1962. Prague: Publishing House of the Czechoslovak Academy of Sciences 1962.

— Die Entwicklung des Diencephalons im Lichte neuer Forschung. Progr. Br. Res. **5**, 223—229 (1964).

—, and B. KÄLLÉN: Studies on the topography of the migration areas in the vertebrate brain. Acta anat. (Basel) **17**, 353—369 (1953a).

— — On the development of neuromeres to migration areas in the vertebrate cerebral tube. Acta anat. (Basel) **18**, 65—73 (1953b).

— — Notes on the early histogenesis and morphogenesis of the central nervous system in vertebrates. J. comp. Neurol. **100**, 627—660 (1954).

— — The archencephalic neuromery in Amblystoma punctatum. An experimental study. Acta anat. (Basel) **24**, 208—214 (1955).

BOTERENBROOD, E. C.: On pattern formation in the prosencephalon; an investigation on desaggregated and reaggregated presumptive prosencephalic material of neurulae of *Triturus alpestris*. Thesis, Utrecht (1962).

BRADLEY, O. C.: Neuromeres of the rhombencephalon of the pig. Rev. Neurol. Psychiat. (Paris) 2, 625—635 (1904).

— On the development of the hind-brain of the pig. J. Anat. Physiol. (Lond.) **40**, 1—14 (1906).

BROMAN, I.: Beschreibung eines Menschlichen Embryos von beinahe 3 mm Länge mit spezieller Bemerkung über die bei demselben befindlichen Hirnfalten. Morph. Arb. **5**, 169—205 (1896).

BUTCHER, E. O.: The development of the somites in the white rat (Mus norvegicus albinus) and the fate of the myotomes, neural tube, and gut in the tail. Amer. J. Anat. **44**, 381—439 (1929).

CHILD, C. M.: Patterns and problems of development. Chicago, Ill.: Chicago University Press 1941.

COGHILL, G. E.: Correlated anatomical and physiological studies of the growth of the nervous system of amphibia. IV. Rates of proliferation and differentiation in the central nervous system of Amblystoma. J. comp. Neurol. **37**, 71—120 (1924).

DELSMAN, H. C.: The ancestry of vertebrates. Amersfoort: Valkoff & Co. 1922.

DOHRN, A.: Der Ursprung der Wirbelthiere und das Prinzip des Functionswechsels. Genealogische Skizzen, Leipzig 1875.

EYAL-GILADI, H.: Dynamic aspects of neural induction in Amphibia. Arch. Biol. Liège et Paris **65**, 179—259 (1954).

FOSTER, M., and F. M. BALFOUR: Grundzüge der Entwicklungsgeschichte der Thiere. Leipzig 1876.

FRORIEP, A.: Bemerkungen zur Frage nach der Wirbeltheorie des Kopfskelettes. Anat. Anz. **2**, 815—835 (1887).

— Einige Bemerkungen zur Kopffrage. Anat. Anz. **21**, 545—553 (1902).

FUJITA, S.: Kinetics of cellular proliferation. Exp. Cell Res. **28**, 52—60 (1962).

—, and S. MIYAKE: Selective labeling of cell groups and its application to cell identification. Exp. Cell Res. **28**, 158—161 (1962).

GEGENBAUR, C.: Die Metamerie des Kopfes und die Wirbeltheorie des Kopfskelettes. Morph. Jb. **13**, 1—114 (1887).

GOETTE, A.: Die Entwicklungsgeschichte der Unke (Bombinator igneus). Leipzig: Leopold Voss 1875.

GOODRICH, E. S.: Metameric segmentation and homology. Quart. J. micr. Sci. **59**, 227—248 (1913).

— On the development of the segments of the head in Scyllium. Quart. J. micr. Sci. **63**, 1—30 (1918).

GORONOWITSCH, N.: Untersuchungen über die Entwickelung der sogenannten „Ganglienleisten" im Kopfe der Vogelembryonen. Morph. Jb. **20**, 187—259 (1893).

GROENBERG, G.: Die Ontogenese eines niederen Säugergehirns nach Untersuchungen an Erinaceus europaeus. Zool. Jb., Abt. Anat. u. Ontog. **15**, 261—384 (1901).

GRÜNTHAL, E.: Untersuchungen zur Ontogenese und über den Bauplan des Gehirns. In: K. FEREMUTSCH u. E. GRÜNTHAL, Beiträge zur Entwicklungsgeschichte und normalen Anatomie des Gehirns. Bibl. psychiat. neurol. (Basel) **91**, 5—32 (1952).

HALLER, G.: Die Gliederung des Zwischen- und Mittelhirns der Wirbeltiere. Morph. Jb. **63**, 359—407 (1929).

HAMBURGER, V., and H. L. HAMILTON: A series of normal stages in the development of the chick embryo. J. Morph. 88, 49—92 (1951).

HARA, K.: Regional neural differentiation inducted by prechordal and presumptive chordal mesoderm in the chick embryo. Thesis, Utrecht (1961).

HARKMARK, W., and K. GRAHAM: A method for opening and closing eggs with an experimental demonstration of its advantages. Anat. Rec. **110**, 41—48 (1951).

HERRICK, C. J.: The internal structure of the midbrain and thalamus of Necturus. J. comp. Neurol. 28, 215—348 (1917).

— The brain of the Tiger salamander. Chicago, Ill.: Chicago University Press 1948.

Hertwig, O.: Handbuch der vergleichenden und experimentellen Entwicklungslehre der Wirbeltiere, Bd. 2, Teil 3. Jena: Gustav Fischer 1906.
Hill, C.: Primary segments of the vertebrate head. Anat. Anz. **16**, 535—369 (1899).
— Developmental history of primary segments of the vertebrate head. Zool. Jb., Abt. Anat. u. Ontog. **13**, 393—446 (1900).
Hillemann, H. H.: An experimental study of the development of the pituitary gland in chick embryos. J. exp. Morph. **3**, 347—374 (1943).
Hinsch, G. W., and H. L. Hamilton: The developmental fate of the first somite of the chick. Anat. Rec. **125**, 225—245 (1956).
His, W.: Ueber die erste Anlage des Wirbelthierleibes. Verh. naturforsch. Ges. Basel **4** (1868).
— Ueber die Gliederung des Gehirns. Verh. naturforsch. Ges. Basel **5** (1873).
— Anatomie menschlicher Embryonen. I. Embryonen der ersten Monate. Leipzig: F. C. W. Vogel 1880.
— Zur Geschichte des Gehirns sowie der centralen und peripherischen Nervenbahnen beim Menschlichen Embryo. Abh. K. S. Ges. Wiss. **24**, 341—392 (1888).
— Zur allgemeinen Morphologie des Gehirns. Arch. Anat. Physiol., Anat. Abt. 346—383 (1892).
Hochstetter, F.: Beiträge zur Entwicklungsgeschichte des menschlichen Gehirns, Teil 1. Wien u. Leipzig 1919.
Hoffmann, C. K.: Über die Metamerie des Nachhirns und Hinterhirns, und ihre Beziehung zu den segmentalen Kopfnerven bei Reptilienembryonen. Zool. Anz. **12**, 337—339 (1889).
Holmdahl, D. E.: Die erste Entwicklung des Körpers bei den Vögeln und Säugetieren, inkl. dem Menschen, besonders mit Rücksicht auf die Bildung des Rückenmarks, des Zöloms und der entodermalen Kloake nebst einem Exkurs über die Entstehung der Spina bifida in der Lumbosakralregion. Morph. Jb. **54**, 333—384 (1925).
— Die Entstehung und weitere Entwicklung der Neuralleiste (Ganglienleiste) bei Vögeln und Säugetieren. Z. mikr.-anat. Forsch. **14**, 99—298 (1928).
Hubbard, M. E.: Some experiments on the order of succession of the somites of the chick. Amer. Naturalist **42**, 466—471 (1908).
Hughes, W. L., V. P. Bond, G. Brecher, E. P. Cronkite, R. B. Painter, H. Quastler, and F. G. Sherman: Cellular proliferation in the mouse as revealed by autoradiography with tritiated thymidine. Proc. nat. Acad. Sci. (Wash.) **44**, 476—483 (1958).
Hugosson, R.: Morphologic and experimental studies on the development and significance of the rhombencephalic longitudinal cell columns. Thesis, Lund, Sweden: Håkon Ohlssons boktryckeri 1957.
Ingall, N. W.: Beschreibung eines menschlichen Embryo von 4,9 mm. Arch. mikr. Anat. **70**, 505—576 (1907).
Jager, J.: Über die Segmentierung der Hinterhauptregion und die Beziehung der Cartilago acrochordalis zur Mesodermcommissur. Morph. Jb. **56**, 1—21 (1926).
Johnston, J. B.: The morphology of the forebrain vesicle in vertebrates. J. comp. Neurol. **19**, 457—539 (1909).
— Notes on the neuromeres of the brain and spinal cord. Anat. Rec. **10**, 209—210 (1916).
Källén, B.: The nuclear development in the mammalian forebrain with special regard to the subpallium. Kungl. Fysiogr. Sällsk. Lund Förh. **61**, Nr 9, 10 p. (1951a)
— Embryological studies on the nuclei and their homologization in the vertebrate forebrain. Kungl. Fysiogr. Sällsk. Lund Förh. **62**, Nr 11, 36 p. (1951b).
— Notes on the proliferation processes in the neuromeres in vertebrate embryos. Acta Soc. Med. upsalien. **57**, 111—118 (1952).
— On the significance of the neuromeres and similar structures in vertebrate embryos. J. Embryl. exp. Morph. **1**, 387—392 (1953).
— On the segmentation of the central nervous system. Kungl. Fysiogr. Sällsk. Lund Förh. **64**, Nr 18, 10 p. (1954).
— Neuromery in living and fixed chick embryos. Acta Univ. lund., Avd. 2, **25**, Nr 5, 6 p. (1955).
— Studies on the mitotic activity in chick and rabbit brains during ontogenesis. Acta Univ. lund., Avd. 2, **26**, Nr. 17, 14 p. (1956a).

KÄLLÉN, B.: Experiments on neuromery in Ambystoma punctatum embryos. J. Embryol. exp. Morph. **4**, 66—72 (1956b).

— Contribution to the knowledge of the regulation of the proliferation processes in the vertebrate brain during ontogenesis. Acta anat. (Basel) **27**, 351—360 (1956c).

— Embryol. aspects of the concept of homology. Arch. Zool. **12**, 137—142 (1959).

— Early morphogenesis and pattern formation in the central nervous system. In: Organogenesis, p. 107—128. New York: Holt Reinehart and Winston 1965.

—, and B. LINDSKOG: Formation and disappearance of neuromery in Mus musculus. Acta anat. (Basel) **18**, 273—282 (1953).

KAMON, H.: Zur Entwicklungsgeschichte des Gehirns des Hühnchens. Anat. H., Abt. 1 **30**, 560—650 (1906).

KINGSBURY, B. F.: The extent of the floor-plate of His and its significance. J. comp. Neurol. **32**, 113—137 (1920).

— The fundamental plan of the vertebrate brain. J. comp. Neurol. **34**, 461—492 (1922).

— The developmental significance of the notochord (Chorda dorsalis). Z. Morph. Anthr. **24**, 59—74 (1924).

— Branchiomerism and the theory of head segmentation. J. Morph. **42**, 83—109 (1926).

— The "law" of cephalo-caudal differential growth in its application to the nervous system. J. comp. Neurol. **56**, 431—463 (1932).

— The development of the septum medulla. J. comp. Neurol. **60**, 81—110 (1934).

KÖLLIKER, A. v.: Entwicklungsgeschichte des Menschen und der höheren Thiere. Leipzig 1861.

KOLLMANN, J.: Die Entwicklung der Chorda dorsalis bei dem Menschen. Anat. Anz. **5**, 308—321 (1890).

KUHLENBECK, H.: Über die morphologische Bewertung der sekundären Neuromerie. Anat. Anz. **81**, 129—148 (1935).

— The ontogenetic development of the diencephalic centres in a bird's brain (chick). J. comp. Neurol. **66**, 23—76 (1937).

— The ontogenetic development and phylogenetic significance of the cortex telencephali in the chick. J. comp. Neurol. **69**, 273—302 (1938).

— The human diencephalon. A summary of development, structure, function and pathology. Confin. neurol. (Basel) **14** Suppl., 1—230 (1954).

— Die Formbestandteile der Regio pretectalis des Anamnier-Gehirns und ihre Beziehungen zum Hirnbauplan. Folia anat. jap. **28**, 23—44 (1956).

KUPFFER, C. v.: Studien zur vergleichenden Entwicklungsgeschichte des Kopfes der Kranioten (1895).

— Die Morphogenesis des Centralnervensystems. In: Handbuch der vergleichenden und experimentellen Entwickelungslehre der Wirbeltiere (ed. O. HERTWIG), Bd. 2 Teil 3. Jena: Gustav Fischer 1906.

LEHMANN, F. E.: Einführung in die physiologische Embryologie. Basel: Birkhäuser 1945.

LILLIE, F. R.: The development of the chick. An introduction to embryology. New York: Henry Holt & Co. 1919.

LOCY, W. A. C.: Contribution to the structure and development of the vertebrate head. J. Morph. **11**, 497—594 (1895).

MALPIGHI, M.: Diss. epistolica de formatione pulli in ovo. London 1673. Cit. from F. Tiedemann, 1816.

MARSHALL, A. M.: Development of the cranial nerves in the chick. Quart. J. micr. Sci. **18**, 10—40 (1878).

MARTIN, A., and J. LANGMAN: The development of the spinal cord examined by autoradiography. J. Embryol. exp. Morph. **14**, 25—35 (1965).

MCCLURE, C. F. W.: The segmentation of the primitive vertebrate brain. J. Morph. **4**, 35—56 (1890).

MCQUADE, H. A., M. FRIEDKIN, and A. ATCHISON: Chromosome abbreviations caused by radioactive thymidine. Nature (Lond.) **175**, 1038—1039 (1955).

MEEK, A.: The segments of the vertebrate brain and head. Anat. Anz. **31**, 408—415 (1907).

— The encephalomeres and cranial nerves of an embryo of Acanthias vulgaris. Anat. Anz. **34**, 473—475 (1909).

Meek, A.: The cranial segments and nerves of the rabbit with some remarks on the phylogeny of the nervous system. Anat. Anz. **36**, 605—572 (1910).
Menkens, B.: Experiments on somite-differentiation. Slow-motion pictures presented at the VIII Internat. Embryological Conference, Interlaken 1967.
Messier, B., and C. P. Leblond: Preparation of coated radioautographs by dipping sections in fluid emulsion. Proc. Soc. exp. Biol. (N.Y.) **96**, 7—10 (1957).
Mihalkovics, V. v.: Entwicklungsgeschichte des Gehirns. Leipzig 1877.
Minot, C. S.: Human embryology. Boston, Mass.: Copyr. S. Minot 1892.
Neal, H. V.: The segmentation of the nervous system in Squalus acanthiasis. Bull. Mus. comp. Zool. Harv. **31**, 293—315 (1898).
— Neuromeres and metameres. J. Morph. **31**, 293—315 (1918).
Neumayer, L.: Studie zur Entwicklungsgeschichte des Gehirns der Säugetiere. Festschr. z. siebenzigsten Geburtst. v. C. von Kupffer, Teil I—II, S. 455—486. Jena 1899.
Nieuwkoop, P. D.: Induction and pattern formation as primary mechanisms in early embryonic differentiation. Wageningen, international lecture course "Cell differentiation and morphogenesis", April 26th—29th (1965).
— E. C. Boterenbrood, A. Kremer, F. F. S. N. Bloesma, E. L. M. J. Hoessels, S. Meyer, and F. J. Verheyen: Activation and organization of the central nervous system in amphibians. I, II and III. J. exp. Zool. **120**, 1—108 (1952).
— — Origin and establishment of organization pattern in embryonic fields during early development in amphibians and birds, in particular in the nervous system and its substrate. Proc. kon. ned. Akad. Wet., Ser. C **58**, 219—227 (Part I), 356—367 (Part II) (1955).
Orr, H.: Contribution to the embryology of the Lizard. J. Morph. **1**, 311—372 (1887).
Palmgren, A.: Embryological and morphological studies on the mid-brain and cerebellum of vertebrates. Acta zool. (Stockh.) **2**, 1—94 (1921).
Patten, B. M.: Foundations of embryology. London: McGraw-Hill Book Co. 1958.
Patterson, J. T.: The order of appearance of the anterior somites in the chick. Biol. Bull. mar. biol. Lab., Woods Hole **13**, 121—133 (1907).
Pearse, A. G. E.: Histochemistry, 2nd ed. London: J. & A. Churchill 1960.
Platt, J. B.: Further contribution to the morphology of the vertebrate head. Anat. Anz. **6**, 251—265 (1891).
Rendahl, H.: Embryologische und Morphologische Studien über das Zwischenhirn beim Huhn. Acta zool. (Stockh.) **5**, 241—344 (1924).
Romanoff, A. L.: The avian embryo. New York: The MacMillan Company 1960.
Rugh, R.: Experimental embryology. Minneapolis: Burgess Publ. Co. 1948.
Šala, M.: Distribution of activating and transforming influences in amphibians. Proc. kon. ned. Akad. Wet., Ser. C **58**, 632—647 (1955).
Sanders, F. K.: Cytochemical differentiation between the pentose and desoxypentose nucleic acids in tissue sections. Quart. J. micr. Sci. **87**, 203—208 (1946).
Saxén, L., and S. Toivonen: Primary embryonic induction. London: Academic Press 1962.
Schumacher, O.: Beiträge zur Entwicklungsgeschichte des Vertebratengehirns. IV. Die Entwicklungsgeschichte des Kiebitzgehirns. Z. Anat. Entwickl.-Gesch. **87**, 139—251 (1928).
Scott, W. B.: Notes on the development of Petromyzon. J. Morph. **1**, 253—310 (1887).
Sidman, R. L., I. L. Miale, and Ned Feder: Cell migration in the immature neuro-epithelium of mouse embryos studied by autoradiography with tritiated thymidine. Anat. Rec. **133**, 450 (1959a).
— — — Cell proliferation and migration in the primitive ependymal zone. Exp. Neurol. **1**, 322—333 (1959b).
Spemann, H.: Über die Determination der ersten Organanlagen des Amphibienembryo I—IV. Arch. Entwickl.-Mech. Org. **43**, 448—555 (1918).
Spratt, N. T., Jr.: Regression and shortening of the primitive streak in the explanted chick blastoderm. J. exp. Zool. **104**, 69—100 (1947).
— Localization of the prospective neural plate in the early chick blastoderm. J. exp. Zool. **120**, 109—130 (1952).
Streeter, G. L.: The nuclei of origin of the cranial nerves in the 10 mm human embryo. Anat. Rec. **2**, 111—115 (1908).

STREETER, G. L.: Die Entwicklung des Nervensystems. In: Handbuch der Entwicklungsgeschichte des Menschen (ed. F. KEIBEL and F. P. MALL), Bd. 2, S. 1—156. Leipzig: S. Hirzel 1911.
— The status of metamerism in the central nervous system of chick embryos. J. comp. Neurol. **57**, 455 (1933).
SUNDBERG, C.: Das Glykogen in menschlichen Embryonen von 15, 27 and 40 mm. Z. Anat. Entwickl.-Gesch. **73**, 168—246 (1924).
THOMPSON, P.: Description of a human embryo of twenty-three paired somites. J. Anat. Physiol. (Lond.) **41**, 159—171 (1907).
TIEDEMANN, F.: Anatomie und Bildungsgeschichte des Gehirns im Foetus des Menschen. Nürnberg 1816.
VAAGE, S.: A study of the cell migrations in mesencephalon and metencephalon in chick embryos by autoradiography. Rep. Fourth Scand. Congr. Cell Res., p. 50 (1965).
— Histogenesis of the nuclei Isthmi in the chick. An autoradiographic study. (1969, in preparation.)
—, and H. O. HØIVIK: Experimental investigations on the neuromeres and the occipital somites in chick embryos. (1969, in preparation.)
VEIT, O.: Beiträge zur Kenntnis des Kopfes der Wirbeltiere. III. Morph. Jb. **84**, 86—107 (1939).
VERSLUYS, J.: Über die Rückbildung der Kiemenbögen bei den Selachii. Bijdragen tot de Dierkunde. Feest-nummer, uitgegeven bij gelegenheit van den 70 sten Geboortedag van Dr. MAX WEBER, S. 95—105. Leiden: E. J. Brill 1922.
VIRGILIO, G. DI, N. LAVENDA, and J. L. WORDEN: Sequence of events in neural tube closure and the formation of neural crest in the chick embryo. Acta anat. (Basel) **68**, 127—146 (1967).
WATERS, B. W.: Primitive segmentation of the vertebrate brain. Quart. J. micr. Sci. **33**, 457—476 (1892).
WEBER, A.: Contribution a l'etude de la metamerie du cerveau anterieur chez quelques oiseaux. Arch. Anat. micr. Morph. exp. **3**, 369—423 (1900).
WEDIN, B.: Proneuromeres and neuromeres in the cerebral tube of Torpedo ocellata. Acta anat. (Basel) **21**, 59—69 (1954).
WILLIAMS, L. W.: The somites of the chick. Amer. J. Anat. **11**, 55—100 (1910).
ZIEGLER, H. E.: Die phylogenetische Entstehung des Kopfes der Wirbeltiere. Jena. Z. Med. Naturw. **43**, 653—684 (1908).
ZIEHEN, T.: Morphogenie des Centralnervensystems der Säugetiere. In: Handbuch der vergleichenden und experimentellen Entwicklungslehre der Wirbeltiere (ed. O. HERTWIG), Bd. 2, Teil 3. Berlin 1906.
ZIMMERMANN, W.: Über die Metamerie des Wirbeltierkopfes. Verh. anat. Ges. **5**, 107—113 (1891).

Index